T0134702

Mathematical Theories of Machine Learning - Theory and Applications

Bin Shi • S. S. Iyengar

Mathematical Theories of Machine Learning - Theory and Applications

Springer

Bin Shi
University of California
Berkeley, USA

S. S. Iyengar
Florida International University
Miami, FL, USA

ISBN 978-3-030-17078-3 ISBN 978-3-030-17076-9 (eBook)
https://doi.org/10.1007/978-3-030-17076-9

This Springer imprint is published by the registered company Springer Nature Switzerland AG.
The registered company address is: Gewerbestrasse 11, 6330 Cham, Switzerland

Dedicated to Prof. Tao Li (Late) for his contributions to Machine Learning and AI and Prof. Ruzena Bajcsy (UC Berkeley) for her life-long contribution to Robotics and AI

Foreword

The field of machine learning will be significantly impacted by this book. While there have been several books that address the different classes of machine learning techniques, this book examines the mathematical foundation for machine learning algorithms in depth. This is needed because practitioners and academics must have a way to measure the effectiveness of the large number of algorithms against applications.

One fundamental contribution the authors discuss is convexly constrained-sparse subspace clustering (CoCoSSC). Several of the machine learning techniques depend on convergence of a steepest descent approach. The CoCoSSC method permits designing of a faster convergence for the gradient descent technique when the objective function requires some elements of nonconvex objectives (or convexly constrained objectives).

There are many applications that would benefit from this foundational work. Machine learning applied to cyber security is one such application. There are several practical applications where the goal is to reduce the amount of data overwhelming the cyber analysts. Specifically, there are examples where a logistic regression classifier, based on steepest gradient descent, helps in the separation of relevant cyber topics from non-cyber topics in a corpus of data. Another similar application is in identifying exploited malware that is a subset of a large vulnerability database.

Furthermore, artificial intelligence has the potential to revolutionize many industries, for example, applications ranging from driverless cars, finance, national security, medicine, and e-commerce, to name a few. This book applies to these types of applications by advancing the understanding of the mathematics behind convex and constrained optimization techniques as it applies to steepest gradient descent for optimization, which is fundamental to several classes of machine learning algorithms.

Boston, MA David R. Martinez

Preface

> Machine learning is a core, transformative way by which we're rethinking everything we're doing. We're thoughtfully applying it across all our products, be it search, ads, YouTube, or Play. We're in the early days, but you'll see us in a systematic way think about how we can apply machine learning to all these areas.
>
> —Sundar Pichai, CEO, Google

One of the most interesting topics of research with the potential to change the way the world is headed is machine learning and the associated techniques. However, in the current state of the art, the machine learning research does not have a solid theoretical framework that could form the basis for the analysis and provide guidelines for the experimental runs. This book is an attempt to identify and address the existing issues in the respective field of great research interest in the modern outlook on machine learning, artificial intelligence, deep neural networks, etc. For all the great wonders that these abovementioned techniques can do, it is still a mystery as to how to use the basic concepts they so highly depend on. Gradient descent is one of the popular techniques that has been widely deployed in training any neural network. One of the challenges that erupts while using gradient descent is the absence of guidelines on when they converge, be it to local or a global minima. In this book, we have attempted to address this crucial problem. This book offers to the readers novel theoretical frameworks that could be used in analyzing the convergence behavior.

This book also represents a major contribution in terms of mathematical aspects of machine learning by the authors and collaborators. Throughout the book, we have made sure the reader gets a good understanding and feel of the theoretical frameworks that are and can be employed in the gradient descent technique and the ways of deploying them in the training of neural networks. To emphasize this, we have used results from some of our recent research along with a blend of what is being explored by other researchers. As the readers read through the chapters of the book, they would be exposed to the various applications of great importance, like the subspace clustering and time series analysis. This book thus tries to strike a balance in the way theory is presented along with some of the applications that come hand in hand with it. Through this book, we hope to make the reading more exciting and also have a huge impact on the readers by providing them with the right tools in the machine learning domain.

In drawing comparisons to existing books like the one titled "*Deep Learning*" by Goodfellow, Bengio, and Courville, this book digs deep into defining and showcasing the latest research in the fields of gradient descent that makes it a more comprehensive tool for students and professionals alike. Also, the emphasis given to relating the concepts to applications like subspace clustering and time series data makes it a better alternative to most other books in this field.

The intended audience for this book includes anyone who is actively working in machine learning, be it students, professors, industry professionals, or even independent researchers. We have tried to compile the book to provide a handy handbook for day-to-day activities.

The book is organized into many free-flowing parts so that the reader is first exposed to the basic concepts of machine learning, neural networks, optimization, gradient descent, etc. In the following parts and sections, the reader can study and understand the optimality and adaptivity of choosing step sizes of gradient descent and thus escaping the strict saddle points in the non-convex optimization problems. We first show for a fixed step size the maximum allowable step size for gradient descent to find a local minimizer, which is twice the inverse of gradient Lipschitz constant ($1/L$) when all the saddle points are strict. However, since the gradient descent with (fixed) step sizes exceeding $2/L$ diverges for worst-case functions, we obtain the optimal step size of gradient descent for strict saddle non-convex optimization problems. The main observation is that as long as the induced mapping by gradient descent is a local diffeomorphism, the points that converge to any strict saddle point have Lebesgue measure "0," whereas previous works all required this mapping to be a global diffeomorphism. We also consider adaptive choices of step sizes and show that if the step size at each iteration is proportional to the inverse of the local gradient Lipschitz constant, gradient descent will not converge to any strict saddle point. To our knowledge, this is the first result showing gradient descent with varying step sizes can also escape saddle points. This is proved by applying a generalization of Hartman product map Theorem from dynamical systems theory.

The subsequent parts of the book define and elaborate on algorithms used in finding the local minima in non-convex optimization scenarios and thus aid in obtaining the global minima that pertains in some degree to Newton's second law without friction. With the key observation of the velocity observable and controllable in the motion, the algorithms simulate Newton's second law without friction based on the symplectic Euler scheme. From the intuitive analysis of analytical solutions, we give a theoretical analysis for the high-speed convergence in the proposed algorithm. Finally, experiments for strongly convex, non-strongly convex, and non-convex functions in higher dimensions are proposed. This book also describes the implementation of the discrete strategies that would be used in testing the observability and controllability of velocity, or kinetic energy, as well as in the artificial dissipation of energies.

This part is followed by the study of the problem subspace clustering with noisy and missing data—a problem well-motivated by practical applications. Applications consider data subject to stochastic Gaussian noise and/or incomplete data with uniformly missing entries. Our main contribution is the development of CoCoSSC, a novel noisy subspace clustering method inspired by CoCoLasso. Notably, CoCoSSC uses a semi-definite programming-based preprocessing step to "de-bias" and "denoise" the input data before passing into the Lasso SSC algorithm, which makes it significantly more robust and an l1-normalized self-regression program. We prove theoretical guarantees that CoCoSSC works even when an overwhelming $1 - \Omega\left(n^{-2/5}\right)$ fraction of the data is missing and when the data is contaminated by an additive Gaussian noise with a vanishing signal-to-noise ratio (SnR) of $n^{-1/4}$. These rates significantly improve over what is previously known, which only handle a constant fraction of missing data and an SnR of $n^{-1/6}$ for Gaussian noise. Compared to existing methods in the literature, our approach enjoys improved sample completive inference strategy of particle learning. Extensive empirical studies on both the synthetic and real application time series data are conducted to demonstrate the effectiveness and the efficiency of the introduced methodology, and computationally efficient numerical results confirmed the effectiveness and efficiency of our proposed method.

Berkeley, USA | Bin Shi
Miami, FL, USA | S. S. Iyengar

Acknowledgments

The authors are very grateful to Professor Tao Li for his contributions to machine learning and for his directions and advise in the earlier stages of compilation of the content of this book. The authors also thank all collaborators from Carnegie Mellon University; University of California, Santa Barbara; University of Miami; and University of Southern California for all their invaluable inputs and comments which have motivated the authors toward the successful completion of the book. More importantly, the authors would like to thank Dr. Yining Wang, Jason Lee, Simon S. Du, Yuxiang Wang, Yudong Tau, and Wentao Wang for their comments and contributions on this book. Many of these results have been submitted for publication to various conferences and journals. So, we want to thank these publishers for their support.

The authors would also like to extend their warm gratitude toward all other collaborating faculty from FIU notably Dr. Kianoosh G. Boroojeni, and graduate students like Sanjeev Kaushik Ramani, among others, for all the help and support they provided toward the successful completion of this book. The authors are also thankful to the IBM Yorktown Heights, Xerox, National Science Foundation, and other agencies for providing the necessary funds used to carry out this research. Majority of the work comes from Bi Shi's research work.

Contents

Author Biographies

Dr. Bin Shi is currently a postdoctoral researcher at the University of California at Berkeley. His preliminary research focus was on the theory of machine learning, specifically on optimization. Bin received his B.Sc. in Applied Math from Ocean University of China in 2006 followed by M.Sc. in Pure Mathematics from Fudan University, China, in 2011 and M.Sc. in Theoretical Physics from the University of Massachusetts, Dartmouth. Bin's research interests focus on statistical machine learning and optimization, and some theoretical computer science. Bin's research contributions have been published in NIPS OPT-2017 Workshop and *INFORMS Journal on Optimization*—Special Issue for Machine Learning.

Dr. S. S. Iyengar is a Distinguished University Professor and Distinguished Ryder Professor and Director of the School of Computing and Information Sciences at Florida International University, Miami. Dr. Iyengar is a pioneer in the field of distributed sensor networks/sensor fusion, computational aspects of robotics, and high-performance computing.

He was a visiting Satish Dhawan Professor at IISC, Bangalore, and also Homi Bhabha Professor at IGCAR, Kalpakkam, Tamil Nadu. He was a visiting professor at the University of Paris, Tsinghua University, KAIST, etc.

He has published over 600 research papers and has authored/edited 22 books published by MIT Press, John Wiley & Sons, Prentice Hall, CRC Press, Springer Verlag, etc. These publications have been used in major universities all over the world. He has many patents and some patents are featured in the World's Best Technology Forum in Dallas, Texas. His research publications are on the design and analysis of efficient algorithms, parallel computing, sensor networks, and robotics. During the last four decades, he has supervised over 55 Ph.D. students, 100 master's students, and many undergraduate students who are now faculty at major universities worldwide or scientists or engineers at national labs/industries around the world. He has also had many undergraduate students working on his research projects. Recently, Dr. Iyengar received the Times Network 2017 Nonresident Indian of the Year Award, a prestigious award for global Indian leaders.

Dr. Iyengar is a member of the European Academy of Sciences, a Fellow of IEEE, a Fellow of ACM, a Fellow of AAAS, a Fellow of the National Academy of Inventors (NAI), a Fellow of Society of Design and Process Program (SPDS), a Fellow of Institution of Engineers (FIE), and a Fellow of the American Institute for Medical and Biological Engineering (AIMBE). He was awarded a Distinguished Alumnus Award of the Indian Institute of Science, Bangalore, and the IEEE Computer Society Technical Achievement for the contributions to sensor fusion algorithms and parallel algorithms. He also received the IBM Distinguished Faculty Award and NASA Fellowship Summer Award at Jet Propulsion Laboratory. He is a Village Fellow of the Academy of Transdisciplinary Learning and Advanced Studies in Austin, Texas, 2010.

He has received various national and international awards including the Times Network NRI (Nonresident Indian) of the Year Award for 2017, the National Academy of Inventors Fellow Award in 2013, the NRI Mahatma Gandhi Pravasi Medal at the House of Lords in London in 2013, and a Lifetime Achievement Award conferred by the International Society of Agile Manufacturing (ISAM) in recognition of his illustrious career in teaching, research, and administration and a lifelong contribution to the fields of Engineering and Computer Science at Indian Institute of Technology (BHU). In 2012, Iyengar and Nulogix were awarded the 2012 Innovation-2-Industry (i2i) Florida Award. Iyengar received a Distinguished Research Award from Xiamen University, China, for his research in sensor networks, computer vision, and image processing. Iyengar's landmark contributions with his research group include the development of grid coverage for surveillance and target location in distributed sensor networks and the Brooks–Iyengar fusion algorithm. He was a recipient of a Fulbright Distinguished Research Award; 2019 IEEE Intelligence and Security Informatics Research Leadership Award; Lifetime Achievement Award for his contribution to Distributed Sensor Networks at the 25th International IEEE High Performance Computing Conference (2019), which was given by Dr. Narayana Murthy (co-founder of Infosys); Florida Innovation Award to Industry for innovation technology for glaucoma device; LSU Rainmaker Award; and Distinguished Research Master Award. He has also been awarded Honorary and Doctorate of Science and Engineering Degree. He serves on the advisory board of many corporations and universities around the world. He has served on many national science boards such as NIH's National Library of Medicine in Bioinformatics, National Science Foundation review panel, NASA Space Science, Department of Homeland Security, Office of Naval Security, and many others. His contribution to the US Naval Research Laboratory was a centerpiece of a pioneering effort to develop image analysis for science and technology and to expand the goals of the US Naval Research Laboratory.

The impact of his research contributions can be seen in companies and national labs like Raytheon, Telcordia, Motorola, the United States Navy, DARPA, and other US agencies. His contribution in DARPA's program demonstration with BBN, Cambridge, Massachusetts, MURI, researchers from PSU/ARL, Duke, University of Wisconsin, UCLA, Cornell University, and LSU has been significant. He is also the founding editor of the *International Journal of Distributed Sensor Networks*. He has been on the editorial board of many journals and is also a PhD committee member at various universities, including CMU, Duke University, and many others throughout the world. He is presently the editor of *ACM Computing Surveys* and other journals.

He is also the founding director of the FIU's Discovery Laboratory. His research work has been cited extensively. His fundamental work has been transitioned into unique technologies. All through his four-decade long professional career, Dr. Iyengar has devoted and employed mathematical morphology in a unique way for quantitative understanding of computational processes for many applications.

Part I
Introduction

Chapter 1
Introduction

Learning has various definitions based on the context in which it is used and the various entities involved in the learning process. The need for machines to learn and thus adapt to the changes in its surroundings led to the rise of the field aptly called "machine learning." A machine is expected to learn and predict the future outcomes based on the changes that it notices in the external structure, data/inputs fed that would have to be responded to, and the program/function it was built to perform. This forms the basis for the various complex computational capabilities that any modern artificially intelligent based (AI-based) system would require and includes computations dealing with recognition of patterns, diagnosis of data, controlling the system or environment, planning activities, etc.

Machine learning (ML) as a field of study has evolved over time owing to the merging and binding of the concepts and methods defined and specified mainly in statistics and artificial intelligence (AI) along with other budding and related areas. One of the important characteristic traits of ML is the ease with which the algorithms that are designed and defined can handle very complex data sets and provide accurate predictions and help in deciphering new knowledge. There is however a very fundamental question as to reason for the need of such predictive models which requires the machine to learn and become more efficient in tackling day-to-day situations in which it is exposed. Some of the compelling reasons for this are as listed below:

- Humans update their knowledge of the surroundings on a constant basis and thus are able to make appropriate decisions for the various situations to which they are exposed. Letting the machine understand its surroundings and thus adapt to the situation in similar lines would aid them in enhancing their decision-making capabilities.
- The newer technologies have led to the generation and thus the explosion of data. However, it should be noted that most of these data sets are highly unstructured, thus making it difficult to design and devise a program that can handle it

B. Shi, S. S. Iyengar, *Mathematical Theories of Machine Learning - Theory and Applications*, https://doi.org/10.1007/978-3-030-17076-9_1

efficiently. Having a machine that understands and learns from the input and previous outputs fed as training data prevents the need for redesigns.
- It is difficult to predict the environment in which machines would be working when it is being manufactured and the cost of having a custom-made machine for performing specific tasks is very high. This also creates a strong case for the need for machines that can adapt, modify, and alter the way they work making the integration into the society and legacy systems seamless.
- Identifying relationships among data of varying kinds also plays a vital role in the predictions. In some situations, it is difficult for even humans to crunch all the data in identifying such relationships. Machine learning algorithms provide the ideal foil in identification of the relationships and thus the abstraction and prediction of possible outcomes.

Along with the dependence of ML algorithms on statistics and AI, there are various other theories and models that have seamlessly converged and aided in making ML more robust and widened the horizons to which ML is applicable. Some of these fields are as listed below:

- **Advances in Control Theoretic Sciences:** Working with unknown parameters would need better estimation something that can be derived from the concepts in control theory. The system should also be able to adapt and track the changes that happen during the processing.
- **Cognitive and psychological models:** The learning rate and the efficiency of learning vary from one human to another and studying these changes and the basis for these changes paves the way to design more robust algorithms suitable for the application. These aspects are used when working with ML and the derived applications.
- **Theory of Evolution:** The well-known evolution theory worked towards defining and identifying the way humans and other species evolved also provided necessary inputs and directions in the design of ML algorithms for various applications. Also the development of the brain and the activities help in defining better neural networks and deep learning algorithms.

1.1 Neural Networks

Works by many researchers in this field have brought to light the various benefits of interconnecting non-linear elements and networks of such elements with weights that can be altered to have a major impact on the way ML algorithms work. Networks designed in this intention form the neural networks. Neural networks by themselves have found a lot of applications and have been a topic that has interested researchers for a long time now. The parallels that can be drawn between these

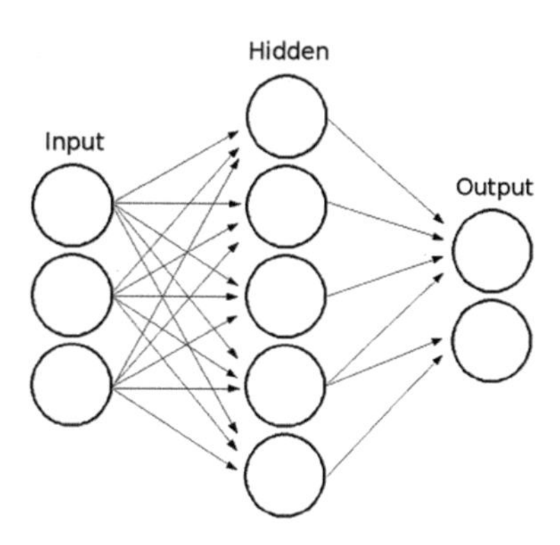

Fig. 1.1 Deep neural network

networks and the actual biological nervous systems with its associated neurons and their network make it easier in designing and identifying solutions to the various problems.

Neural networks consist of multiple layers of computational units and elements similar to the neurons in the human nervous system, which perform the necessary computations. The computation results in the transformation of the raw input into expected or predictable outputs for further use. Each of these units (also known as neurons) multiplies the input it receives with a certain predefined alterable weight, performs the computation, and forwards the results to the next layer. The neurons aggregate various results from the previous layer neurons, adjust the weight and the bias, normalize the result using an activation function, and then use it as the input. This is depicted in Fig. 1.1.

1.1.1 Learning Process That Is Iterative

One of the salient features of using neural networks in the application is the way in which the learning happens in an iterative manner with the neurons at each layer performing and adapting based on the inputs it receives from the previous layers. It also adjusts the weights to be applied along with neutralizing or adjusting the bias and the normalization of the values. This happens for every iteration and is a part of the "*learning phase.*" The network would try some possible values to identify the best fit, then uses it to train the input samples.

There have been many identified advantages and applications of using neural networks. Some of the important advantages include the high fault tolerance they possess towards corrupted data inputs and the ease with which they adapt to newer data that is different from the training data sets. With all these advantages, there are many popular neural network algorithms in use in our day-to-day computation. One of the popular algorithms from that list is the "*backpropagation algorithm.*"

The initial phase of using a neural network in an application includes the building of a training data set and using it in the training process. This is the learning phase where the network and thus the system learns the various input types and samples and classifies, then categorizes them into known samples which can be used when working with actual data sets. There are initial weights that would be applied to the data while they are being manipulated. Some of these weights are randomly chosen initially and refined and altered as the learning process goes on. The training set would pose certain expected outputs that are compared to each level in the hidden layers and this leads towards getting a more accurate value for the multiplicative weights.

There are various known applications of neural networks. In this chapter of the book, we have selected a list of applications with which to introduce the concept and the plethora of applications. We begin with convolutional neural networks.

A. Convolutional Neural Networks Convolutional neural networks are also called CNN or ConvNets that are a specialized category of the neural networks which uses convolution in the computation. The use of convolution helps in the separation and thus the extraction of certain features and thus the CNN has many applications in the field of image processing and computer vision. Feature extractions help in pattern identification in applications that span a very wide range starting from the face recognition feature in the modern smartphones to even the navigation modules of the ambitious projects of many companies in building autonomous cars.

Specialized learning and adequate training data sets should help the applications to identify the region of interest (ROI) and the remaining portions. There are various means of accelerating the process and thus harnessing better performance. The hidden layers with their capabilities provide the CNN with more versatile options that can improve the fault tolerance when using such algorithms.

B. Recurrent Neural Networks An important application of using neural networks is in the processing of data sequences and thus predicting the outcomes in the future. This is supported by the recurrent neural networks, sometimes addressed as an RNN. RNNs are very popular for applications that deal with natural language processing. RNNs use sequential data which is an improvised version of the traditional neural networks that assume the data at each layer to be independent. The way RNN works is by performing the same set of operations/functions on all the input elements recursively and thus the output is highly dependent on the previous inputs. This is one of the neural network variants that has memory with which to work with and can handle very long sequences.

RNN assumes that the predictions that are expected as an output are probabilistic. A typical RNN is as shown in Fig. 1.2. Another example apart from the usage in NLP could be to use it in the prediction of how a stock would fare in the market on a certain day.

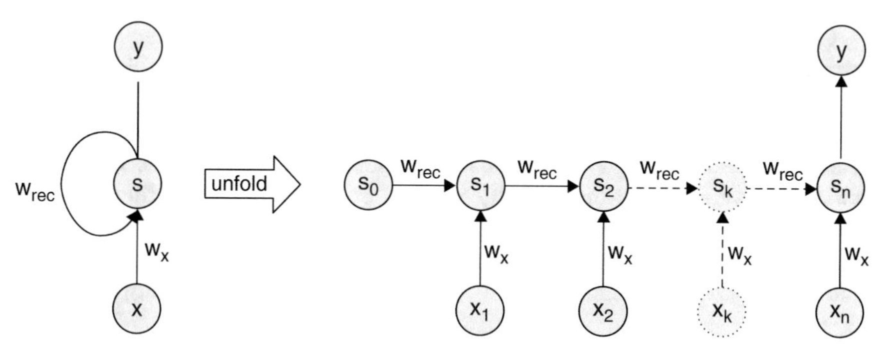

Fig. 1.2 RNN

Fig. 1.3 Deep neural networks as a sub-category of ML

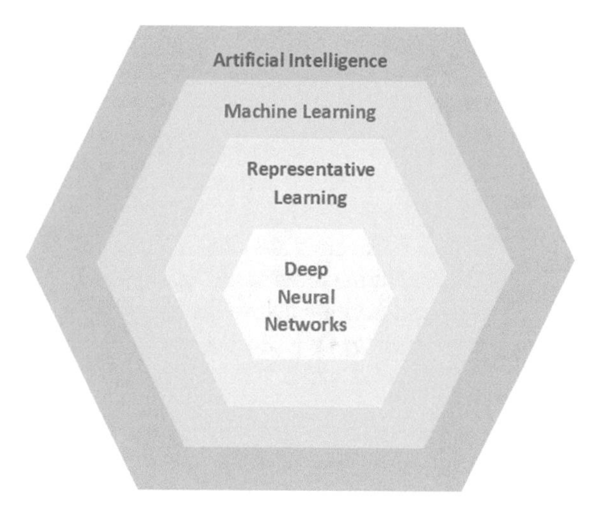

1.2 Deep Learning

Another subset of machine learning that has found many applications and the attention of researchers is deep learning. It is a subset of ML and the derived methods that help in designing and structuring the processing of data based on the way a human brain would function. It involves many concepts of neural networks that were defined in the previous subsection (Fig. 1.3).

The input data is processed in ways and means a biological brain would process and try to modify and transform the input into representations that are highly abstract and aggregated. Taking an example of the image processing application, the raw image or video stream sent as an input in the various layers get processed differently and then gets encoded optimally so that the learning process is complete and thus the system can be adapted to predicting or performing more complex operations on newer data sets.

Certain applications of deep learning include modification of audio in films, extraction of edges and thus in processing images and video streams, classification of objects and enhancement of them in older photographs. Some other applications include handwriting detections and analysis and many other newer applications with recommendation systems and sentiment analysis modules. Also, the application horizons have extended dramatically when it is associated with artificial intelligence. Machine learning algorithms can be "supervised" or "unsupervised," that is if the output classes are known, then the learning algorithm is supervised, otherwise it is not. Supervised machine learning is based on the statistical learning theory.

1.3 Gradient Descent

By far one of the major optimization strategies currently used in machine learning and deep learning by researchers is gradient descent (GD). In this book, we discuss some novel techniques that would help us better understand and use this concept in visualizing many more applications. It is a concept that has a major impact during the training phase of designing the ML algorithm. GD is based on a convex function whose parameters are altered in an increasing manner that would lead to the minimization of the provided function till the local minima is achieved. On the whole, GD is an algorithm that works towards minimizing the given function.

Based on the input they process and the way they work, the gradient descent algorithms are categorized as follows:

1.3.1 Batch Gradient Descent

It has also been referred to as "vanilla" (pure) gradient descent for its simple, small steps towards the minima, works by calculating the error for each of the input data sets during the training period. It incorporates a form of cycle processing which leads to consistent model updates. Since it happens at the training phase, it is called a training epoch.

It has its own set of advantages. One of the most important is the way in which it enhances the efficiency and works towards the stabilization of the errors and leads to faster convergence. Batch gradient descent however has some disadvantages where the earlier stabilized error convergence value would sometimes not be the best but would let an overall convergence occur. Another major disadvantage is the need for a memory component without which this algorithm would not work.

1.3.2 Stochastic Gradient Descent

Another category of gradient descent is known as stochastic gradient descent (SGD), wherein the process is performed for each of the example inputs. The updates happen one at a time and for each of the inputs. Thus, the SGD is faster than the BGD and there is a pretty detailed rate of improvement. However, frequent updates on each of the inputs make it a computationally expensive process. However, depending on the criticality of the application and the need, the frequency of updates can be altered.

1.3.3 Mini-Batch Gradient Descent

A preferred hybrid algorithm most often used by researchers is the mini-batch gradient descent. In this technique the data set is split into smaller batches and updates happen for each of these batches. It thus taps into the advantages of both the above categories while addressing some of the disadvantages discussed earlier. The batch size and the frequency of updates both can be altered making it the most famous and common type of gradient descent used in many of the applications.

1.4 Summary

On the whole, dealing with solutions utilizing machine learning involves the formulation of the problem by defining the various parameters involved and providing details. These details play an important role in defining the models and algorithms that would be trained using the training data sets, thereby leading to the appropriate classifying, predicting, and clustering of the data.

Once the initial training is complete, more complex data sets closer to the application of use would be fed in as input for the training process. Creating the ideal training and test data sets is crucial in determining the model performance when exposed to real-life situations and scenarios. Another important aspect that affects the functioning of the model (or algorithm) is the availability of necessary resources for computation.

It should be noted that training, classification, prediction, and clustering are all computationally intensive tasks due to the large data sets required. Thus, analysis of these large volumes consumes appreciable amounts of computational capability. The presence of all the above would lead to the seamless, effective, and efficient working of the devised ML algorithm (Fig. 1.4).

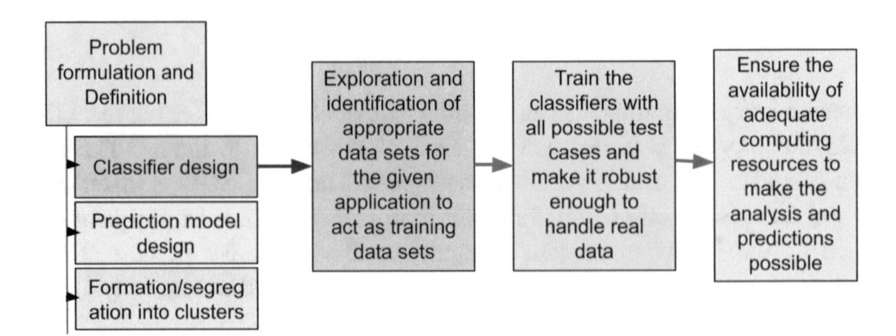

Fig. 1.4 Overview of ML phases

1.5 Organization of the Research Monograph

The machine learning research currently runs as an experimental exercise with no solid theoretical analysis and guidelines. There is a critical need for a theoretical framework. In this context, the proposed book will be a welcome tool for students and professionals alike.

This book is organized into many parts containing chapters that will enthuse the reader into understanding the optimality and adaptivity of choosing step sizes of gradient descent and thus escaping the strict saddle points in the non-convex optimization problems. In Part I, we introduce the basic concepts along with references and description of the framework and problems that would be discussed in this book. The observations pertaining to the experiments and derived concepts are explained in great detail. The variations that occur with adaptive choices of step sizes are also discussed. One of the important observations is that the step size at each iteration is proportional to the inverse of the local gradient Lipschitz constant, gradient descent will not converge to any strict saddle point. This is the first result of which we are aware demonstrating that a gradient descent with varying step size can also escape saddle points. This is proved by applying a generalization of Hartman product map theorem from dynamical systems theory.

Part II of the book defines and elaborates on algorithms used in finding the local minima in non-convex optimization scenarios and thus aid in obtaining the global minima. From the intuitive analysis of analytical solution, we give a theoretical analysis for the high-speed convergence in the algorithm proposed.

Finally, we develop procedures for demonstrating strongly convex, non-strongly convex, and non-convex functions in these higher-dimensions. This book also describes the implementation of the discrete strategies that would be used in testing how one can observe and control velocity, or kinetic energy, as well as the artificial dissipation of energies.

In the last part, we introduce a novel VAR model with elastic-net regularization and its equivalent Bayesian model allowing for both a stable sparsity and a group selection, and a time-varying hidden interaction discovery among multiple time series. We develop a solution capable of capturing the time-varying temporal dependency structure through Bayesian modeling and particle learning's fully adaptive inference strategy. We conduct extensive empirical studies on both the synthetic and real application time-series data as we demonstrate the introduced method's effectiveness and the efficiency.

Chapter 2
General Framework of Mathematics

With the explosive growth of data nowadays, a young and interdisciplinary field, *data science*, has emerged, which uses scientific methods, processes, algorithms and systems to extract knowledge and insights from data in various forms, both structured and unstructured. This data science field is becoming popular and needs to be developed urgently so that it can serve and guide for the industry of the society. Rigorously, applied *data science* is a "concept to unify statistics, data analysis, machine learning and their related methods" in order to "understand and analyze actual phenomena" with data. It employs techniques and theories drawn from many fields within the context of mathematics, statistics, information science, and computer science.

Within the field of data analytics, *machine learning* is a method used to devise complex models and algorithms that lend themselves to prediction; in commercial use, this is known as predictive analytics. The name *machine learning* was coined in 1959 by Arthur Samuel, which evolved from the study of pattern recognition and computational learning theory in artificial intelligence. *Computational statistics*, which also focuses on prediction-making through the use of computers, is a closely related field and often overlaps with *machine learning*.

The name, *computational statistics*, implies that it is composed of two indispensable parts, statistics inference models and the corresponding algorithms implemented in computers. Based on the different kinds of hypotheses, statistics inference can be divided into two schools, frequentist inference school and Bayesian inference school. Here, we describe each one briefly. Let $\mathcal{P}$ be a premise and $\mathcal{O}$ be an observation which may give evidence for $\mathcal{P}$. The priori $P(\mathcal{P})$ is the probability that $\mathcal{P}$ is true before the observation is considered. Also, the posterior $P(\mathcal{P}|\mathcal{O})$ is the probability that $\mathcal{P}$ is true after the observation $\mathcal{O}$ is considered. The likelihood $P(\mathcal{O}|\mathcal{P})$ is the chance of observation $\mathcal{O}$ when evidence $\mathcal{P}$ exists. Finally, $P(\mathcal{O})$ is the total probability, calculated in the following way:

B. Shi, S. S. Iyengar, *Mathematical Theories of Machine Learning - Theory and Applications*, https://doi.org/10.1007/978-3-030-17076-9_2

$$P(\mathcal{O}) = \sum_{\mathcal{P}} P(\mathcal{O}|\mathcal{P}) P(\mathcal{P}).$$

Connecting the probabilities above is the significant Bayes' formula in the theory of probability

$$P(\mathcal{P}|\mathcal{O}) = \frac{P(\mathcal{O}|\mathcal{P}) P(\mathcal{P})}{P(\mathcal{O})} \sim P(\mathcal{O}|\mathcal{P}) P(\mathcal{P}), \tag{2.1}$$

where $P(\mathcal{O})$ can be calculated automatically if we have known the likelihood $P(\mathcal{O}|\mathcal{P})$ and $P(\mathcal{P})$. If we presume that some hypothesis (parameter specifying the conditional distribution of the data) is true and that the observed data is sampled from that distribution, that is,

$$P(\mathcal{P}) = 1,$$

only using conditional distributions of data given the specific hypotheses are the view of the frequentist school. However, if there is no presumption that some hypothesis (parameter specifying the conditional distribution of the data) is true, that is, there is a prior probability for the hypothesis $\mathcal{P}$,

$$\mathcal{P} \sim P(\mathcal{P}),$$

summing up the information from the prior and likelihood is the view from the Bayesian school. Apparently, the view from the frequentist school is a special case of the view from the Bayesian school, but the view from the Bayesian school is more comprehensive and requires more information.

Take the Gaussian distribution with known variance for the likelihood as an example. Without loss of generality, we assume the variance $\sigma^2 = 1$. In other words, the data point is viewed as a random variable $\mathbf{X}$ following the rule below:

$$\mathbf{X} \sim P(x|\mathcal{P}) = \frac{1}{\sqrt{2\pi}} e^{-\frac{(x-\mu)^2}{2}},$$

where the hypothesis is $\mathcal{P} = \{\mu | \mu \in (-\infty, \infty)$ is some fixed real number$\}$. Let the data set be $\mathcal{O} = \{x_i\}_{i=1}^n$. The frequentist school requires us to compute maximum likelihood or maximum log-likelihood, that is,

$$\begin{aligned}
\underset{\mu\in(-\infty,\infty)}{\operatorname{argmax}} f(\mu) &= \underset{\mu\in(-\infty,\infty)}{\operatorname{argmax}} \log P(\mathcal{O}|\mathcal{P}) \\
&= \underset{\mu\in(-\infty,\infty)}{\operatorname{argmax}} \left(\log \prod_{i=1}^{n} P(x_i \in \mathcal{O}|\mathcal{P})\right) \\
&= \underset{\mu\in(-\infty,\infty)}{\operatorname{argmax}} \log\left[\left(\frac{1}{\sqrt{2\pi}}\right)^{n} e^{-\frac{\sum_{i=1}^{n}(x_i-\mu)^2}{2}}\right] \\
&= -\underset{\mu\in(-\infty,\infty)}{\operatorname{argmin}} \left[\frac{1}{2}\sum_{i=1}^{n}(x_i-\mu)^2 + n\log\sqrt{2\pi}\right],
\end{aligned} \tag{2.2}$$

which has been shown in the classical textbooks, such as [RS15], whereas the Bayesian school requires to compute maximum posterior estimate or maximum log-posterior estimate, that is, we need to assume reasonable prior distribution.

- If the prior distribution is a Gauss distribution $\mu \sim \mathcal{N}(0, \sigma_0^2)$, we have

$$\begin{aligned}
&\underset{\mu\in(-\infty,\infty)}{\operatorname{argmax}} f(\mu) \\
&= \underset{\mu\in(-\infty,\infty)}{\operatorname{argmax}} \log P(\mathcal{O}|\mathcal{P})\, P(\mathcal{P}) \\
&= \underset{\mu\in(-\infty,\infty)}{\operatorname{argmax}} \log\left(\prod_{i=1}^{n} \log P(x_i \in \mathcal{O}|\mathcal{P})\right) P(\mathcal{P}) \\
&= \underset{\mu\in(-\infty,\infty)}{\operatorname{argmax}} \log\left\{\left[\left(\frac{1}{\sqrt{2\pi}}\right)^{n} e^{-\frac{\sum_{i=1}^{n}(x_i-\mu)^2}{2}}\right]\cdot\left(\frac{1}{\sqrt{2\pi}\,\sigma_0}\right) e^{-\frac{\mu^2}{2\sigma_0^2}}\right\} \\
&= -\underset{\mu\in(-\infty,\infty)}{\operatorname{argmin}} \left[\frac{1}{2}\sum_{i=1}^{n}(x_i-\mu)^2 + \frac{1}{2\sigma_0^2}\cdot\mu^2 + n\log\sqrt{2\pi} + \log\sqrt{2\pi}\,\sigma_0\right].
\end{aligned} \tag{2.3}$$

- If the prior distribution is a Laplace distribution $\mu \sim \mathcal{L}(0, \sigma_0^2)$, we have

$$\begin{aligned}
&\max_{\mu\in(-\infty,\infty)} f(\mu) \\
&= \underset{\mu\in(-\infty,\infty)}{\operatorname{argmax}} \log P(\mathcal{O}|\mathcal{P})\, P(\mathcal{P}) \\
&= \underset{\mu\in(-\infty,\infty)}{\operatorname{argmax}} \log\left(\prod_{i=1}^{n} \log P(x_i \in \mathcal{O}|\mathcal{P})\right) P(\mathcal{P}) \\
&= \underset{\mu\in(-\infty,\infty)}{\operatorname{argmax}} \log\left\{\left[\left(\frac{1}{\sqrt{2\pi}}\right)^{n} e^{-\frac{\sum_{i=1}^{n}(x_i-\mu)^2}{2}}\right]\cdot\left(\frac{1}{2\sigma_0^2}\right) e^{-\frac{|\mu|}{\sigma_0^2}}\right\} \\
&= -\underset{\mu\in(-\infty,\infty)}{\operatorname{argmin}} \left[\frac{1}{2}\sum_{i=1}^{n}(x_i-\mu)^2 + \frac{1}{\sigma_0^2}\cdot|\mu| + n\log\sqrt{2\pi} + \log 2\sigma_0^2\right].
\end{aligned} \tag{2.4}$$

- If the prior distribution is the mixed distribution combined with Laplace distribution and Gaussian distribution $\mu \sim \mathcal{M}(0, \sigma_{0,1}^2, \sigma_{0,2}^2)$, we have

$$
\begin{aligned}
&\underset{\mu\in(-\infty,\infty)}{\operatorname{argmax}} \; f(\mu) \\
&= \underset{\mu\in(-\infty,\infty)}{\operatorname{argmax}} \; \log P(\mathcal{O}|\mathcal{P})\, P(\mathcal{P}) \\
&= \underset{\mu\in(-\infty,\infty)}{\operatorname{argmax}} \; \log\left(\prod_{i=1}^{n} \log P(x_i \in \mathcal{O}|\mathcal{P})\right) P(\mathcal{P}) \\
&= \underset{\mu\in(-\infty,\infty)}{\operatorname{argmax}} \; \log\left\{\left[\left(\frac{1}{\sqrt{2\pi}}\right)^n e^{-\frac{\sum_{i=1}^{n}(x_i-\mu)^2}{2}}\right]\right. \\
&\qquad\qquad \left. \cdot\, C(\sigma_{0,1}, \sigma_{0,2})^{-1} e^{-\frac{|\mu|}{\sigma_{0,1}^2} - \frac{\mu^2}{2\sigma_{0,2}^2}}\right\} \\
&= -\underset{\mu\in(-\infty,\infty)}{\operatorname{argmin}} \left[\frac{1}{2}\sum_{i=1}^{n}(x_i-\mu)^2 + \frac{1}{\sigma_0^2}\cdot|\mu| + \frac{1}{2\sigma_{0,2}^2}\cdot\mu^2\right. \\
&\qquad\qquad \left. + n\log\sqrt{2\pi} + \log C(\sigma_{0,1}, \sigma_{0,2})\right],
\end{aligned}
\tag{2.5}
$$

where $C = 2\sqrt{2\pi}\sigma_{0,1}^2\sigma_{0,2}$.

2.1 Concluding Remarks

In summary, based on the description in this chapter, a statistical problem can be solved by transforming it into an optimization problem. The required proof to validate this statement was outlined and provided in this chapter. In the following chapter we discuss the problem further by identifying how it is formulated and we develop an approach to tackle the problem.

Chapter 3
Optimization Formulation

Based on the description on the statistics model in the previous section, we formulate the problems that we need to solve from two angles. One is from the field of optimization, the other is from samples of probability distribution. Practically, from the view of efficient algorithms in computers, the representation of the first one is the expectation–maximization (**EM**) algorithm. The EM algorithm is used to find (local) maximum likelihood parameters of a statistical model in scenarios wherein the equations cannot be solved directly. These models use latent variables along with unknown parameters and known data observations, i.e., either there is a possibility of finding missing values among the data or the model can be formulated in more simple terms by assuming the existence of unobserved data points. A mixture model can be described in simplistic terms with an assumption that each of the observed data points have a corresponding unobserved data point, or latent variable that specifies the mixture component to which each of the data points belong.

The EM algorithm moves on with the observation that there is a way that these two sets of equations can be solved numerically. The solution starts by picking an arbitrary value for either of the two sets of unknowns, and use this in estimating the other set. The new values are then used in finding better estimates of the first set. This alternation between the two continues until the resulting values converge to fixed points. There is no guarantee that this approach would work but has been proven to be a worthwhile option to try. It is also observed that the derivative of the likelihood is very close to zero at that point, which indicates that the point is either a maximum or a saddle point. In most cases, there is a possibility of occurrence of multiple maximas, with no assurance that the global maximum will be found. Some likelihoods also have singularities in them, i.e., a nonsensical maxima. A solution that may be found by EM in a mixture model involves the setting of one of the components to have zero variance with the mean parameter of the same component equal to one of the data points.

The second one is the Markov chain Monte Carlo (**MCMC**) method. MCMC methods are primarily used for calculation of the numerical approximations of

B. Shi, S. S. Iyengar, *Mathematical Theories of Machine Learning - Theory and Applications*, https://doi.org/10.1007/978-3-030-17076-9_3

multi-dimensional integrals, as used in Bayesian statistics, computational physics, computational biology, and computational linguistics.

3.1 Optimization Techniques Needed for Machine Learning

Recall the process of finding the maximum probability, which is equivalent to the maximum log-likelihood or the maximum log-posterior estimate is essential. We describe them rigorously in statistics language as below.

- Finding the maximum likelihood (2.2) is equivalent to the expression below:

$$\underset{\mu\in(-\infty,\infty)}{\operatorname{argmax}} \ f(\mu) = - \underset{\mu\in(-\infty,\infty)}{\operatorname{argmin}} \left[\frac{1}{2}\sum_{i=1}^{n}(x_i - \mu)^2 \right], \tag{3.1}$$

  which is named **linear regression** in statistics.
- Finding the maximum posterior estimate (2.3) is equivalent to the expression below:

$$\underset{\mu\in(-\infty,\infty)}{\operatorname{argmax}} \ f(\mu) = - \underset{\mu\in(-\infty,\infty)}{\operatorname{argmin}} \left[\frac{1}{2}\sum_{i=1}^{n}(x_i - \mu)^2 + \frac{1}{2\sigma_0^2}\cdot\mu^2 \right], \tag{3.2}$$

  which is named **ridge regression** in statistics.
- Finding the maximum posterior estimate (2.3) is equivalent to the expression below:

$$\underset{\mu\in(-\infty,\infty)}{\operatorname{argmax}} \ f(\mu) = - \underset{\mu\in(-\infty,\infty)}{\operatorname{argmin}} \left[\frac{1}{2}\sum_{i=1}^{n}(x_i - \mu)^2 + \frac{1}{\sigma_0^2}\cdot|\mu| \right], \tag{3.3}$$

  which is named **lasso** in statistics.
- Finding the maximum posterior estimate (2.3) is equivalent to the expression below:

$$\underset{\mu\in(-\infty,\infty)}{\operatorname{argmax}} \ f(\mu) = - \underset{\mu\in(-\infty,\infty)}{\operatorname{argmin}} \left[\frac{1}{2}\sum_{i=1}^{n}(x_i - \mu)^2 + \frac{1}{\sigma_{0,1}^2}\cdot|\mu| + \frac{1}{2\sigma_{0,2}^2}\cdot\mu^2 \right], \tag{3.4}$$

  which is named **elastic-net** in statistics.

Linear regression (3.1) is considered as one of the standard models in statistics, the variants (3.2)–(3.4) of which are viewed as linear regression with regularizers. Every regularizer has its own advantage, the advantage of ridge regression (3.2) is stability, that of lasso (3.3) is sparsity, and that of elastic-net (3.4) owns sparsity and group effect. Especially, due to the sparse property, the lasso (3.3) becomes one of the most significant models in statistics.

The linear regression and its variants above can be reduced to finding a minimizer of the convex objective function without constraint:

$$\min_{x\in\mathbb{R}} f(x),$$

of which the corresponding high-dimensional expression highly concerned in practice is

$$\min_{x\in\mathbb{R}^n} f(x).$$

All of the descriptions above are from the simple likelihood. In biology, the models above are suitable to study for a single species. Take the tigers in China, for example. There are two kinds of tigers in China, Siberian tiger and South China tiger (Fig. 3.1). If we only consider one kind of tigers, Siberian tiger or South China tiger, then we can assume the likelihood is a single Gaussian; but if we consider the total tigers in China, both Siberian tiger and South China tiger, then the likelihood is a superposition of two single Gaussian. The simple sketch in $\mathbb{R}$ is shown in Fig. 3.2. Comparing the left two and the right one in Fig. 3.2, there exist three stationary points, two local maximal points and one local minimal point. In other words, the objective function is non-convex. The classical convex optimization algorithms,

Fig. 3.1 Left: Siberian tiger; right: South China tiger (courtesy of creative commons)

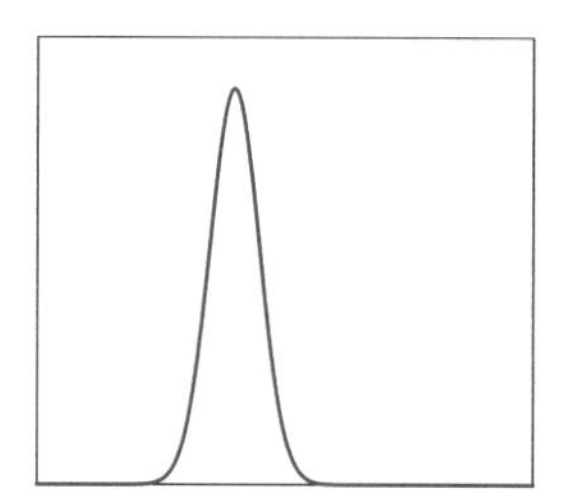
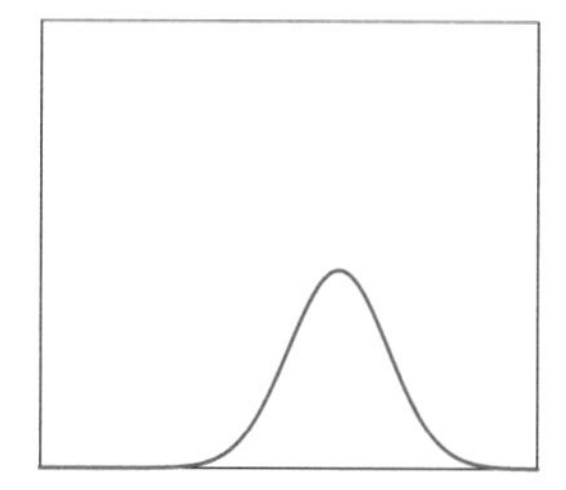
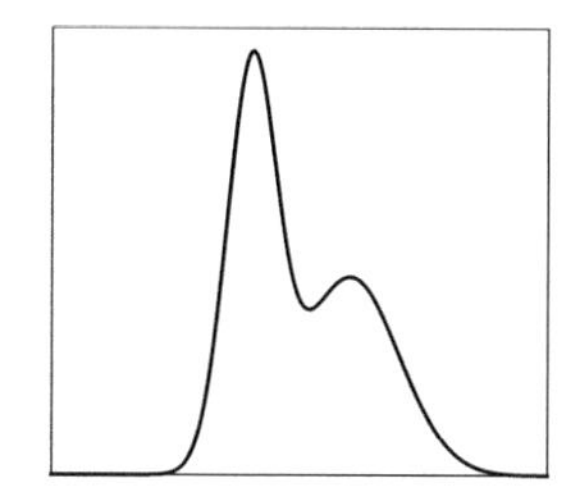

Fig. 3.2 Left: Gaussian-1; middle: Gaussian-2; right: mixed Gaussian: Gaussian-1+Gaussian-2

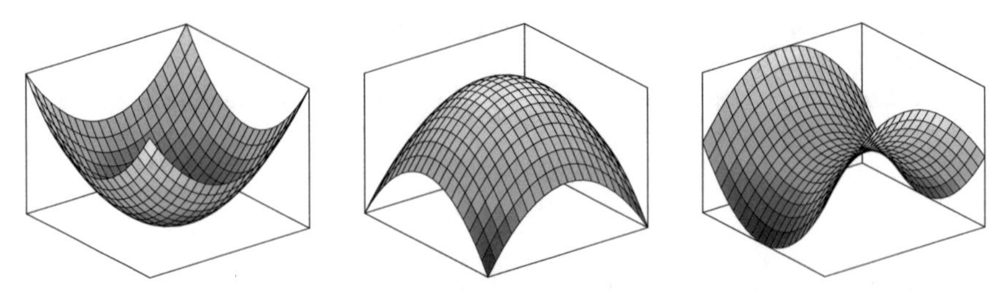

Fig. 3.3 Left: local minimal point; middle: local maximal point; right: saddle

based on the principle that the local minimal point is the global minimal point, are not suitable for the original convex case. Furthermore, if the dimension of the objective function is greater than and equal 2, there exists another stationary point: saddle. We demonstrate the different stationary points in Fig. 3.3.

From the descriptions above, many statistics models are finally transformed to solve an optimization problem, not only simple convex optimization but also complex non-convex optimization. What's more, the optimization algorithms are based on the information from the objective function. The classical oracle assumption for the smoothness is described in [Nes13] as below:

- Zero-order oracle assumption: returns the value $f(x)$;
- First-order oracle assumption: returns the value $f(x)$ and the gradient $\nabla f(x)$;
- Second-order oracle assumption: returns the value $f(x)$, the gradient $\nabla f(x)$, and the Hessian $\nabla^2 f(x)$.

To discriminate if an optimization algorithm is highly efficient in practice, based on the performance, the main characters are from oracle information and iteration complexity. Apparently, zero-order oracle algorithms are firstly considered. Currently, there are two main kinds of methods involved to implement: kernel-based bandit algorithms [BLE17] and algorithms of single-point gradient estimation [FKM05, HL14]. Since the fewer oracle information leads to the higher iteration complexity, the zero-order oracle algorithms are not popular in practice. Furthermore, developing zero-order oracle algorithms is still in the convex stage. Second-order oracle algorithms have been studied widespread for the last four decades, which are essentially based on classical Newton iteration, such as modified Newton's method [MS79], modified Cholesky's method [GM74], cubic-regularization method [NP06], and trust region method [CRS14]. Currently, with the success of deep learning, some algorithms based on Hessian product in non-convex objective have been proposed in [AAZB$^+$17, CD16, CDHS16, LY17, RZS$^+$17, RW17]. However, the difficulty of computing the Hessian information leads to infeasibility in current computers.

Now, we come to the first-order algorithms which have been used widespread. First-order algorithms only need to compute gradient which takes $O(d)$ time complexity, where the dimension d is large. Recall the statistics model (3.1)–(3.4), if we compute the full gradient $\nabla f(\mu)$, it leads to deterministic algorithms; if we compute one gradient $\nabla f_i(\mu)$, that is, $(x_i - \mu)$ for some $1 \leq i \leq n$, it leads to stochastic algorithms. In this research monograph, we focus on deterministic algorithms.

3.1.1 Gradient Descent

Gradient descent (GD) is by far one of the major optimization strategies used in machine learning and deep learning by the researchers currently. In this book, we discuss some novel techniques that would help us better understand and use this concept in visualizing many more applications. It is a concept that has a major impact during the training phase of designing the ML algorithm. GD is based on a convex function whose parameters are altered in an increasing manner that would lead to the minimization of the provided function till the local minima is achieved. On the whole, GD is an algorithm that works towards minimizing the given function.

Machine learning relies on gradient descent (GD) and its many variations as a central optimization methodology in machine learning problems. Given a C^1 or C^2 function $f : \mathbb{R}^n \to \mathbb{R}$ with unconstrained variable $x \in \mathbb{R}^n$, GD uses the following update rule:

$$x_{k+1} = x_k - h_k \nabla f(x_k), \tag{3.5}$$

where h_k are step size, which may be either fixed or vary across iterations. When f is convex, $h_k < \frac{2}{L}$ is a necessary and sufficient condition to guarantee the (worst-case) convergence of GD, where L is the Lipschitz constant of the gradient of the function f.

While this works well for these types of problems, GD is not so well understood in non-convex problems. For general smooth non-convex problems, GD is only known to converge to a stationary point (i.e., a point with zero gradient) [Nes13].

Machine learning tasks often require finding a local minimizer instead of just a stationary point, which can also be a saddle point or a maximizer. Recently, researchers have placed an increasing focus on geometric conditions under which GD escapes saddle points and converges to a local minimizer. More specifically, if the objective function satisfies (1) all saddle point are strict and (2) all local minima are global minima, then GD finds a global optimal solution. These two properties hold for a wide range of machine learning problems, such as matrix factorization [LWL$^+$16], matrix completion [GLM16, GJZ17], matrix sensing [BNS16, PKCS17], tensor decomposition [GHJY15], dictionary learning [SQW17], and phase retrieval [SQW16].

Recent results in this area show that when the objective function has the strict saddle property, then GD converges to a minimizer provided the initialization is randomized and the step sizes are fixed and smaller than $1/L$ [LSJR16, PP16]. While this was the first results establishing convergence of GD, there are still gaps towards fully understanding GD for strict saddle problems.

3.1.2 Accelerated Gradient Descent

Today, many cutting-edge technologies rely on the non-convex optimization algorithm for machine learning, computer vision, natural language processing, and reinforcement learning. Local search methods have become increasingly important, as discovery of a global minimizer in a non-convex optimization problem is NP-hard. These local search methods are all based on the method applied in the convex optimization problem. Formally, the problem of unconstrained optimization is stated in general terms as that of finding the minimum value that a function attains over Euclidean space, i.e.,

$$\min_{x \in \mathbb{R}^n} f(x).$$

Numerous methods and algorithms have been proposed to solve the minimization problem, notably gradient methods, Newton's methods, trust region method, ellipsoid method, and interior-point method [Pol87, Nes13, WN99, LY$^+$84, BV04, B$^+$15].

First-order optimization algorithms have become popular in performing optimization, providing one of the most common methods to optimize neural networks, since the second-order information obtained is supremely expensive. The simplest and earliest method for minimizing a convex function f is the gradient method, i.e.,

$$\begin{cases} x_{k+1} = x_k - h\nabla f(x_k) \\ \text{Any Initial Point}: \; x_0. \end{cases} \tag{3.6}$$

There are two significant improvements of the gradient method to increase the convergence. One of them is the momentum method, which is also known as Polyak heavy ball method, first proposed in [Pol64], i.e.,

$$\begin{cases} x_{k+1} = x_k - h\nabla f(x_k) + \gamma_k(x_k - x_{k-1}) \\ \text{Any Initial Point}: \; x_0. \end{cases} \tag{3.7}$$

Let κ be the condition number, which is the ratio of the smallest eigenvalue and the largest eigenvalue of Hessian at local minima. The momentum method increases the local convergence rate from $1 - 2\kappa$ to $1 - 2\sqrt{\kappa}$. The other is the notorious Nesterov's accelerated gradient method, which was initially proposed in [Nes83] with an improvisation available in [NN88, Nes13], i.e.,

$$\begin{cases} y_{k+1} = x_k - \dfrac{1}{L}\nabla f(x_k) \\ x_{k+1} = x_k + \gamma_k(x_{k+1} - x_k) \\ \text{Any Initial Point}: \; x_0 = y_0, \end{cases} \tag{3.8}$$

where the parameter is set as

$$\gamma_k = \frac{\alpha_k(1-\alpha_k)}{\alpha_k^2 + \alpha_{k+1}} \quad \text{and} \quad \alpha_{k+1}^2 = (1-\alpha_{k+1})\alpha_k^2 + \alpha_{k+1}\kappa.$$

The scheme devised by Nesterov does not only own the property of the local convergence for strongly convex function, but also is the global convergence scheme, from $1 - 2\kappa$ to $1 - \sqrt{\kappa}$ for strongly convex function and from $\mathcal{O}\left(\frac{1}{n}\right)$ to $\mathcal{O}\left(\frac{1}{n^2}\right)$ for non-strongly convex function.

Although there is the complex algebraic trick in Nesterov's accelerated gradient method, the three methods above can be considered from continuous-time limits [Pol64, SBC14, WWJ16, WRJ16] to obtain physical intuition. In other words, the three methods can be regarded as the discrete scheme for solving the ODE. The gradient method as shown in Eq. (3.6) is correspondent to

$$\begin{cases} \dot{x} = -\nabla f(x_k) \\ x(0) = x_0, \end{cases} \tag{3.9}$$

while the momentum method and Nesterov's accelerated gradient method are correspondent to

$$\begin{cases} \ddot{x} + \gamma_t \dot{x} + \nabla f(x) = 0 \\ x(0) = x_0, \ \dot{x}(0) = 0, \end{cases} \tag{3.10}$$

the difference of which are the setting of the friction parameter γ_t. There are two significant intuitive physical meaning in the two ODEs (3.9) and (3.10). The ODE (3.9) is the governing equation for potential flow, a correspondent phenomenon of waterfall from the height along the gradient direction. The infinitesimal generalization is correspondent to heat conduction in nature. Hence, the gradient method (3.6) is viewed as the implement in computer or optimization simulating the phenomena in the real nature. The ODE (3.10) is the governing equation for the heavy ball motion with friction. The infinitesimal generalization is correspondent to chord vibration in nature. Hence, both the momentum method (3.7) and Nesterov's accelerated gradient method (3.8) can be viewed as updated versions of the discrete scheme used to implement and solve by computation or optimization the setting of the friction force parameter.

Furthermore, we can view the three methods above as the thought for dissipating energy implemented in the computer. The unknown objective function in black box model can be viewed as the potential energy. Hence, the initial energy is from the potential function $f(x_0)$ at x_0 to the minimization value $f(x^\star)$ at $x^\star$. The total energy is combined with the kinetic energy and the potential energy. The key observation in this paper is that we find the kinetic energy, or the velocity, is an observable and controllable variable in the optimization process. In other words, we can compare the velocities in every step to look for local minimum in the computational process or reset them to zero to arrive to artificially dissipate energy.

The governing motion equation in a conservation force is introduced and would be used in this book, for comparison as depicted below:

$$\begin{cases} \ddot{x} = -\nabla f(x) \\ x(0) = x_0, \ \dot{x}(0) = 0. \end{cases} \tag{3.11}$$

The phase space concept usually provides all possible values of position and momentum variables. The governing motion equation in a conservation force field (3.11) can be rewritten as

$$\begin{cases} \dot{x} = v \\ \dot{v} = -\nabla f(x) \\ x(0) = x_0, \ v(0) = 0. \end{cases} \tag{3.12}$$

3.1.3 Application to Sparse Subspace Clustering

Another key problem in machine learning, signal processing, and computer vision research is subspace clustering [Vid11]. Subspace clustering aims at grouping data points into disjoint *clusters* so that data points within each cluster lie near a *low-dimensional linear subspace*. It has found many successful applications in computer vision and machine learning, as many high-dimensional data can be approximated by a union of low-dimensional subspaces. Example data include motion trajectories [CK98], face images [BJ03], network hop counts [EBN12], movie ratings [ZFIM12], and social graphs [JCSX11].

Mathematically, let $\mathbf{X} = (\boldsymbol{x}_1, \cdots, \boldsymbol{x}_N)$ be an $n \times N$ data matrix, where n is the ambient dimension and N is the number of data points. We suppose there are L clusters $\mathcal{S}_1, \cdots, \mathcal{S}_L$, and each column (data point) of $\mathbf{X}$ belongs to exactly one cluster, and cluster $\mathcal{S}_\ell$ has $N_\ell \leq N$ points in $\mathbf{X}$. It is further assumed that data points within each subspace lie approximately on a low-dimensional linear subspace $\mathcal{U}_\ell \subseteq \mathbb{R}^n$ of dimension $d_\ell \ll n$. The question is to recover the clustering of all points in $\mathbf{X}$ without additional supervision.

In the case where data are noiseless (i.e., $\boldsymbol{x}_i \in \mathcal{U}_\ell$ if $\boldsymbol{x}_i$ belongs to cluster $\mathcal{S}_\ell$), the following *sparse subspace clustering* [EV13] approach can be used:

$$\text{SSC}: \quad \boldsymbol{c}_i := \arg \min_{\boldsymbol{c}_i \in \mathbb{R}^{N-1}} \|\boldsymbol{c}_i\|_1 \quad s.t. \quad \boldsymbol{x}_i = \mathbf{X}_{-i}\boldsymbol{c}_i. \tag{3.13}$$

The vectors $\{c_i\}_{i=1}^N$ are usually referred to as the *self-similarity matrix*, or simply *similarity matrix*, with the property that $|c_{ij}|$ being large if x_i and x_j belong to the same cluster and vice versa. Afterwards, spectral clustering methods can be applied on $\{c_i\}_{i=1}^N$ to produce the clustering [EV13].

While the noiseless subspace clustering model is ideal for simplified theoretical analysis, in practice data are almost always corrupted by additional noise. A general formulation for the noisy subspace clustering model is $\mathbf{X} = \mathbf{Y} + \mathbf{Z}$, where $\mathbf{Y} = (y_1, \cdots, y_N)$ is an unknown noiseless data matrix (i.e., $y_i \in \mathcal{U}_\ell$ if y_i belongs to $\mathcal{S}_\ell$) and $\mathbf{Z} = (z_1, \cdots, z_N)$ is a noise matrix such that $z_1, \cdots, z_N$ are independent and $\mathbb{E}[z_i|\mathbf{Y}] = \mathbf{0}$. Only the corrupted data matrix $\mathbf{X}$ is observed. Two important examples can be formulated under this framework:

- **Gaussian noise**: $\{z_i\}$ are i.i.d. Gaussian random variables $\mathcal{N}(\mathbf{0}, \sigma^2/n \cdot \mathbf{I}_{n\times n})$.
- **Missing data**: Let $R_{ij} \in \{0, 1\}$ be random variables indicating whether entry $\mathbf{Y}_{ij}$ is observed, that is, $\mathbf{X}_{ij} = R_{ij}\mathbf{Y}_{ij}/\rho$. The noise matrix $\mathbf{Z}$ can be taken as $\mathbf{Z}_{ij} = (1 - R_{ij}/\rho)\mathbf{Y}_{ij}$, where $\rho > 0$ is a parameter governing the probability of observing an entry, that is, $\Pr[R_{ij} = 1] = \rho$.

Many methods have been proposed to cluster noisy data with subspace clustering [SEC14, WX16, QX15, Sol14]. Existing work can be categorized primarily into two formulations: the Lasso SSC formulation

$$\text{LASSO SSC}: \quad c_i := \arg\min_{c_i \in \mathbb{R}^{N-1}} \|c_i\|_1 + \frac{\lambda}{2}\left\|x_i - \mathbf{X}_{-i}c_i\right\|_2^2, \tag{3.14}$$

which was analyzed in [SEC14, WX16, CJW17], and a de-biased Dantzig selector approach

$$\text{DE-BIASED DANTZIG SELECTOR}: \quad c_i := \arg\min_{c_i \in \mathbb{R}^{N-1}} \|c_i\|_1 + \frac{\lambda}{2}\left\|\widetilde{\boldsymbol{\Sigma}}_{-i}c_i - \tilde{\gamma}_i\right\|_\infty, \tag{3.15}$$

which was proposed in [Sol14] and analyzed for an irrelevant feature setting in [QX15]. Here in Eq. (3.15) the terms $\widetilde{\boldsymbol{\Sigma}}_{-i}$ and $\tilde{\gamma}_i$ are de-biased second-order statistics, defined as $\widetilde{\boldsymbol{\Sigma}}_{-i} = \mathbf{X}_{-i}^\top\mathbf{X}_{-i} - \mathbf{D}$ and $\tilde{\gamma}_i = x_i^\top\mathbf{X}_{-i}$, where $\mathbf{D} = \operatorname{diag}(\mathbb{E}[z_1^\top z_1], \cdots, \mathbb{E}[z_N^\top z_N])$ is a diagonal matrix that approximately de-biases the inner product and is assumed to be known. In particular, in the Gaussian noise model we have $\mathbf{D} = \sigma^2\mathbf{I}$ and in the missing data model we have $\mathbf{D} = (1-\rho)^2/\rho \cdot \operatorname{diag}(\|y_1\|_2^2, \cdots, \|y_N\|_2^2)$ which can be approximated by $\hat{\mathbf{D}} = (1-\rho)^2\operatorname{diag}(\mathbf{X}^\top\mathbf{X})$ computable from corrupted data.

3.2 Online Algorithms: Sequential Updating for Machine Learning

In the context of computer science, online algorithms are used to define a set of algorithm that can be used to process the inputs piece-by-piece in a serial fashion. The order of the input matters so much so that the input gets fed to the algorithm and the entire input is not available at the beginning.

Based on the sampling methods, we here briefly introduce the principle behind the online time-varying algorithms. Let $t \in \{0, 1, 2, \ldots, N\}$ be a discrete finite time set. In every $t \in \{0, 1, 2, \ldots, N\}$, there are always new data being observed, noted as $\mathcal{D}_t$. Recall the Bayesian formula (2.1), at time $t = 0$, with the prior $P(\mathcal{H})$ and likelihood $P(D_0|\mathcal{H})$, we have

$$P(\mathcal{H}|D_0) \sim P(D_0|\mathcal{H})P(\mathcal{H}).$$

At time $t = 1$, we take the posterior $P(\mathcal{H}|D_0)$ at time $t = 0$ as the prior at time $t = 1$ and the likelihood $P(D_1|\mathcal{H}, D_0)$, then the new posterior at $t = 1$ can be calculated as

$$P(\mathcal{H}|D_0, D_1) \sim P(D_1|\mathcal{H}, D_0)P(\mathcal{H}|D_0).$$

By analogy, at time $t = N$, we take the posterior $P(\mathcal{H}|D_0, \ldots, D_{N-1})$ at time $t = N - 1$ as the prior at time $t = N$ and the likelihood $P(D_N|\mathcal{H}, D_0, \ldots, D_{N-1})$, then the new posterior at $t = 1$ can be calculated as

$$P(\mathcal{H}|D_0, \ldots, D_N) \sim P(D_N|\mathcal{H}, D_0, \ldots, D_{N-1})P(\mathcal{H}|D_0, \ldots, D_{N-1}).$$

With the description above, we actually implement $N + 1$ times maximum posterior estimate, that is, maximum posterior estimate sequence as below:

$$P(\mathcal{H}|D_0), P(\mathcal{H}|D_0, D_1), \ldots, P(\mathcal{H}|D_0, D_1, D_N).$$

In other words, obtaining the distribution $P(\mathcal{H}|D_0, \ldots, D_k)$ $(k = 0, \ldots, N)$ is sequential updating. With the probability distribution $P(\mathcal{H}|D_0, \ldots, D_k)$ at time $t = k$, we can implement sampling process to generate data to observe the trend from time $t = 0$ to $t = N$ and to compare with the actual trend. Here, without any difficulty, we can find the core part of sequential updating is how to implement the likelihood sequence experimentally

$$P(D_0|\mathcal{H}), P(D_1|\mathcal{H}, D_0), \ldots, P(D_N|\mathcal{H}, D_0, \ldots, D_{N-1}).$$

A popular technique is named as particle learning, which assumes actually the likelihood sequence following Gaussian random walk.

3.2.1 Application to Multivariate Time Series (MTS)

MTS analysis has been extensively employed across diverse application domains [BJRL15, Ham94], such as finance, social network, system management, weather forecast, etc. For example, it is well-known that there exist spatial and temporal correlations between air temperatures across certain regions [JHS+11, BBW+90]. Discovering and quantifying the hidden spatial–temporal dependences of the temperatures at different locations and time brings great benefits for weather forecast, especially in disaster prevention [LZZ+16].

Mining temporal dependency structure from MTS data is extensively studied across diverse domains. The Granger causality framework is the most popular method. The intuition behind it is that if the time series A Granger causes the time series B, the future value prediction of B can be improved by giving the value of A. Regression models have evolved as being one of the principal approaches used for Granger causality. Specifically, in the prediction of the future values of B, one regression model that is built based only on the past values of B should be statistically and significantly less accurate than the regression model inferred by giving the past values of both A and B. Regression model with L_1 regularizer [Tib96], named Lasso-Granger, is an advanced and effective approach for Granger causal relationship analysis. Lasso-Granger can efficiently and effectively help in identifying the sparse Granger Causality especially in high dimensions [BL13].

However, Lasso-Granger suffers some essential disadvantages. The number of non-zero coefficients chosen by Lasso is bounded by the number of training instances and also it tends to randomly select only one variable and ignore the others within a variable group which leads to instability. Moreover, all the work described above assumes a constant dependency structure among MTS. However, this assumption rarely holds in practice, since real-world problems often involve underlying processes that are dynamically evolving over time. Take a scenario in temperature forecast as an example. Local temperature is usually impacted by its neighborhoods, but the dependency relationships dynamically change when monsoon comes from different directions. In order to capture the dynamic dependency typically happening in practice, a hidden Markov regression model [LKJ09] and a time-varying dynamic Bayesian network algorithm [ZWW+16] have been proposed. However, both methods infer the underlying dependency structure based on the offline mode.

3.3 Concluding Remarks

In this chapter, we explicitly model the dynamic changes of the underlying temporal dependencies and infer the model in an online manner.

All the work described earlier is offline, which capture a static relationship. Based on the online model [ZWML16] for the context-aware recommendation, the online Lasso-Granger model is presented in [ZWW$^+$16] for multivariate time series. Compared to the offline approaches, the online Lasso-Granger model can capture the time-varying dependency from multivariate time series. The excellent effects on simulation and evaluation metrics are shown in [ZWW$^+$16]. However, the Lasso-Granger model depends significantly on the scale coefficient of regularizer. When we implement a continuous shrinkage, the Lasso-Granger model is very unstable and provides no group relationship among variables. In other words, we cannot classify the variables to simplify the model. Based on the contents before, we proposed a time-varying elastic-net-Granger inference model.

In this chapter, we investigate the time-varying elastic-net-Granger inference model, which imposes a mixed L_1 and L_2 regularization penalty on the linear regression coefficients. The Elastic_Net regularizer combines advantages of both lasso and ridge regression. Without loss of capturing sparsity, it can capture group effects of variables and has strong stability. Our online elastic-net-Granger inference model is based on particle learning [CJLP10] to capture the dynamical relationship among the variables. Starting from a Gaussian random walk, the dynamical behaviors of the temporal dependency can be modeled. The fully adaptive inference strategy of particle learning effectively obtains the information of the varying dependency. Different from [ZWW$^+$16], we design our own simulator to generate the variables that own group relationship. Our algorithm for online time-varying Bayesian elastic-net Model demonstrates superior performance, far more than that based on Lasso.

Chapter 4
Development of Novel Techniques of CoCoSSC Method

This chapter provides an introduction to our main contributions concerning the development of the novel methods of CoCoSSC is discussed.

4.1 Research Questions

Question 1: Maximum Allowable Fixed Step Size Recall that for convex optimization by gradient decent with fixed step-size rule $h_k \equiv h$, $h < 2/L$ is both a necessary and a sufficient condition for the convergence of GD. However, for non-convex optimization existing works all required the (fixed) step size to be smaller than $1/L$. Because larger step sizes lead to faster convergence, a natural question is to identify the maximum allowable step size such that GD escapes saddle points. The main technical difficulty to analyze larger step size is that the gradient map

$$g(x) = x - h\nabla f(x)$$

may *not* be a diffeomorphism when $h \geq 1/L$. Thus, techniques used in [LSJR16, PP16] are no longer sufficient.

Here, we take a finer look at the dynamics of GD. Our main observation is that the GD procedure escapes strict saddle points under much weaker conditions than g being a diffeomorphism everywhere. In particular, the probability of GD with random initialization converging to a strict saddle point is 0 provided that

$$g(x_k) = x_k - h_t \nabla f(x_k)$$

is a *local* diffeomorphism at every x_t. We further show that

$$\lambda\left(\{h \in [1/L, 2/L) : \exists t, g(x_k) \text{ is not a local diffeomorphism}\}\right) = 0,$$

B. Shi, S. S. Iyengar, *Mathematical Theories of Machine Learning - Theory and Applications*, https://doi.org/10.1007/978-3-030-17076-9_4

where $\lambda(\cdot)$ is the standard Lebesgue measure on $\mathbb{R}$, meaning that for almost every fixed step size choice in $[1/L, 2/L)$, $g(x_k)$ is a local diffeomorphism for every t. Therefore, if a step size h is chosen uniformly at random from $\left(\frac{2}{L} - \epsilon, \frac{2}{L}\right)$ for any $\epsilon > 0$, GD escapes all strict saddle points and converges to a local minimum. See Sect. 7.3 for the precise statement and Sect. 7.5 for the proof.

Question 2: Analysis of Adaptive Step Sizes Another open question we consider in this paper is to analyze the convergence of GD for non-convex objectives when the step sizes $\{h_t\}$ vary as t evolves. In convex optimization, adaptive step-size rules such as exact or backtracking line search [Nes13] are commonly used in practice to improve convergence, and convergence of GD is guaranteed provided that the adaptively tuned step sizes do not exceed twice the inverse of local gradient Lipschitz constant. On the other hand, in non-convex optimization, whether gradient descent with varying step sizes can escape all strict saddle points is unknown.

Existing techniques [LSJR16, PP16, LPP$^+$17, OW17] cannot solve this question because they relied on the classical stable manifold theorem [Shu13], which requires a fixed gradient map, whereas when step sizes vary, the gradient maps also change across iterations. To deal with this issue, we adopt the powerful Hartman product map theorem [Har71], which gives a finer characterization of local behavior of GD and allows the gradient map to change at every iteration. Based on Hartman product map theorem, we show that as long as the step size at each iteration is proportional to the inverse of the *local* gradient Lipschitz constant, GD still escapes all strict saddle points. To the best of our knowledge, this is the first result establishing convergence to local minima for non-convex gradient descent with varying step sizes.

4.2 Accelerated Gradient Descent

We implement our discrete strategies into algorithms with the utility of the observability and controllability of the velocity, or the kinetic energy, as well as artificially dissipating energy for two directions as below:

- The kinetic energy, or the norm of the velocity, is compared with that in the previous step while searching for the local minima in non-convex function or global minima in convex function. It would be reset to zero until it no longer becomes larger.
- An initial larger velocity $v(0) = v_0$ is implemented at any initial position $x(0) = x_0$ so as to identify the global minima in non-convex function. A ball is implemented with (3.12), the local maximum of the kinetic energy is recorded to discern how many local minima exists along the trajectory. The strategy described above is then implemented in identifying the minimum of all the local minima.

For implementing our thought in practice, we utilize the scheme in the numerical method for the Hamiltonian system, the symplectic Euler method. We remark that a more accuracy version is the Störmer–Verlet method for practice.

4.3 The CoCoSSC Method

We now consider an alternative formulation CoCoSSC to solve the noisy subspace clustering problem, inspired by the CoCoLasso estimator for high-dimensional regression with measurement error [DZ17]. First, a pre-processing step is used that computes $\widetilde{\mathbf{\Sigma}} = \mathbf{X}^T\mathbf{X} - \hat{\mathbf{D}}$ and then finds a matrix belonging to the following set:

$$S := \left\{\mathbf{A} \in \mathbb{R}^{N\times N} : \mathbf{A} \succeq \mathbf{0}\right\} \cap \left\{\mathbf{A} : \left|\mathbf{A}_{jk} - \widetilde{\mathbf{\Sigma}}_{jk}\right| \leq |\mathbf{\Delta}_{jk}|, \forall j,k \in [N]\right\}, \qquad (4.1)$$

where $\mathbf{\Delta} \in \mathbb{R}^{N\times N}$ is an error tolerance matrix to be specified by the data analyst. For Gaussian random noise, all entries in $\mathbf{\Delta}$ can be set to a common parameter, while for the missing data model we recommend setting two different parameters for diagonal and off-diagonal elements in $\mathbf{\Delta}$, as estimation errors of these elements of $\mathbf{A}$ behave differently under the missing data model. We give theoretical guidelines on how to set the parameters in $\mathbf{\Delta}$ in our main theorems, while in practice we observe that setting the elements in $\mathbf{\Delta}$ to be sufficiently large would normally yield good results. Because S in Eq. (4.1) is a convex set, and we will later prove that $S \neq \emptyset$ with high probability, a matrix $\tilde{\mathbf{\Sigma}}_+ \in S$ can be easily found by alternating projection from $\tilde{\mathbf{\Sigma}}$.

For any $\widetilde{\mathbf{\Sigma}}_+ \in S$ and let $\widetilde{\mathbf{\Sigma}}_+ = \widetilde{\mathbf{X}}^T\widetilde{\mathbf{X}}$, where $\widetilde{\mathbf{X}} = (\tilde{\boldsymbol{x}}_1, \cdots, \tilde{\boldsymbol{x}}_N) \in \mathbb{R}^{N\times N}$. Such a decomposition exists because $\widetilde{\mathbf{\Sigma}}_+$ is positive semidefinite. The self-regression vector $\boldsymbol{c}_i$ is then obtained by solving the following (convex) optimization problem:

$$\text{CoCoSSC}: \qquad \boldsymbol{c}_i := \arg\min_{\boldsymbol{c}_i \in \mathbb{R}^{N-1}} \|\boldsymbol{c}_i\|_1 + \frac{\lambda}{2}\left\|\tilde{\boldsymbol{x}}_i - \widetilde{\mathbf{X}}_{-i}\boldsymbol{c}_i\right\|_2^2. \qquad (4.2)$$

Equation (4.2) is an ℓ_1-regularized least squares self-regression problem, with the difference of using $\tilde{\boldsymbol{x}}_i$ and $\tilde{\mathbf{X}}_{-i}$ for self-regression instead of directly using the raw noise-corrupted observations $\boldsymbol{x}_i$ and $\mathbf{X}_{-i}$. This leads to improved sample complexity, as shown in Table 4.1 and our main theorems. On the other hand, CoCoSSC retains the nice structure of Lasso SSC, making it easier to optimize. We further discuss this aspect and other advantages of CoCoSSC in the next section.

Table 4.1 Summary of success conditions with normalized signals $\|y_i\|_2 = 1$

	Gaussian model	Missing data (MD)	MD (random subspaces)
LASSO SSC [SEC14]	$\sigma = O(1)$	–	–
LASSO SSC [WX16]	$\sigma = O(n^{1/6})$	$\rho = \Omega(n^{-1/4})$	$\rho = \Omega(n^{-1/4})$
LASSO SSC [CJW17]	–	$\rho = \Omega(1)$	$\rho = \Omega(1)$
PZF-SSC [TV18]	–	$\rho = \Omega(1)$	$\rho = \Omega(1)$
DE-BIASED DANTZIG [QX15]	$\sigma = O(n^{1/4})$	$\rho = \Omega(n^{-1/3})$	$\rho = \Omega(n^{-1/3})$
CoCoSSC (this paper)	$\sigma = O(n^{1/4})$	$\rho = \Omega(\chi^{2/3} n^{-1/3} + n^{-2/5})$[a]	$\rho = \Omega(n^{-2/5})$[a]

Polynomial dependency on d, $\overline{C}$, $\underline{C}$, and $\log N$ are omitted. In the last line χ is a subspace affinity quantity introduced in Definition 9.3 for the non-uniform semi-random model. χ is always upper bounded by $\sqrt{d}$

[a]If $\|y_i\|_2$ is exactly known, the success condition can be improved to $\rho = \Omega(n^{-1/2})$. See Remark 9.3 for details

4.3.1 *Advantages of CoCoSSC*

The CoCoSSC has the following advantages:

1. Equation (4.2) is easier to optimize, especially compared to the de-biased Dantzig selector approach in Eq. (3.15), because it has a smoothly differentiable objective with an ℓ_1 regularization term. Many existing methods such as ADMM [BPC$^+$11] can be used to obtain fast convergence. We refer the readers to [WX16, Appendix B] for further details on efficient implementation of Eq. (4.2). The pre-processing step Eq. (4.1) can also be efficiently computed using alternating projection, as both sets in Eq. (4.1) are convex. On the other hand, the de-biased Dantzig selector formulation in Eq. (3.15) is usually solved using linear programming [CT05, CT07] and could be very slow as the number of variables is large. Indeed, our empirical results show that the de-biased Dantzig selector is almost 5–10 times slower than both LASSO SSC and CoCoSSC.
2. Equation (4.2) has improved or equal sample complexity in both the Gaussian noise model and the missing data model, compared to LASSO SSC and the de-biased Dantzig selector. This is because a "de-biasing" pre-processing step in Eq. (4.1) is used, and an error tolerance matrix $\mathbf{\Delta}$ with different diagonal and off-diagonal elements is considered to reflect the heterogeneous estimation error in $\mathbf{A}$. Table 4.1 gives an overview of our results comparing them with existing results.

4.4 Online Time-Varying Elastic-Net Algorithm

To overcome the deficiency of Lasso-Granger and capture the dynamical change of causal relationships among MTS, we investigate the Granger causality framework with elastic-net [ZH05], which imposes a mixed L_1 and L_2 regularization penalty on the linear regression. The elastic-net cannot only obtain strongly stable coefficients [SHB16], but also capture grouped effects of variables [SHB16, ZH05]. Furthermore, our approach explicitly models the dynamical change behaviors of the dependency as a set of random walk particles, and utilizes particle learning [CJLP10, ZWW+16] to provide a fully adaptive inference strategy. This strategy allows the model to effectively capture varying dependencies while simultaneously learning latent parameters. Empirical studies on synthetic and real data sets demonstrate the effectiveness of our proposed approach.

4.5 Concluding Remarks

In this chapter we defined a novel method that could overcome the deficiency of Lasso-Granger and capture the dynamical change of causal relationships among MTS. We have also discussed an approach that dynamically changes the behavior of the dependency as a set of random walk particles. Empirical studies on both synthetic and real data sets were demonstrated to show the effectiveness of the proposed approach.

Chapter 5
Necessary Notations of the Proposed Method

We define necessary notations and review important definitions that will be used later in our analysis. Let $C^2(\mathbb{R}^n)$ be the vector space of real-valued twice-continuously differentiable functions. Let ∇ be the gradient operator and ∇^2 be the Hessian operator. Let $\|\cdot\|_2$ be the Euclidean norm in $\mathbb{R}^n$. Let μ be the Lebesgue measure in $\mathbb{R}^n$.

Definition 5.1 (Global Gradient Lipschitz Continuity Condition) $f \in C^2(\mathbb{R}^n)$ satisfies the global gradient Lipschitz continuity condition if there exists a constant $L > 0$ such that

$$\|\nabla f(x_1) - \nabla f(x_2)\|_2 \leq L \|x_1 - x_2\|_2 \qquad \forall x_1, x_2 \in \mathbb{R}^n. \tag{5.1}$$

Definition 5.2 (Global Hessian Lipschitz Continuity Condition) $f \in C^2(\mathbb{R}^n)$ satisfies the global Hessian Lipschitz continuity condition if there exists a constant $K > 0$ such that

$$\left\|\nabla^2 f(x_1) - \nabla^2 f(x_2)\right\|_2 \leq K \|x_1 - x_2\|_2 \qquad \forall x_1, x_2 \in \mathbb{R}^n. \tag{5.2}$$

Intuitively, a twice-continuously differentiable function $f \in C^2(\mathbb{R}^n)$ satisfies the global gradient and Hessian Lipschitz continuity condition if its gradients and Hessians do not change dramatically for any two points in $\mathbb{R}^n$. However, the global Lipschitz constant L for many objective functions that arise in machine learning applications (e.g., $f(x) = x^4$) may be large or even non-existent. To handle such cases, one can use a finer definition of gradient continuity that characterizes the *local* behavior of gradients, especially for non-convex functions. This definition is adopted in many subjects of mathematics, such as in dynamical systems research.

B. Shi, S. S. Iyengar, *Mathematical Theories of Machine Learning - Theory and Applications*, https://doi.org/10.1007/978-3-030-17076-9_5

Let $\delta > 0$ be some fixed constant. For every $x_0 \in \mathbb{R}^n$, its δ-closed neighborhood is defined as

$$V(x_0, \delta) = \left\{ x \in \mathbb{R}^n \middle| \, \|x - x_0\|_2 < \delta \right\}. \tag{5.3}$$

Definition 5.3 (Local Gradient Lipschitz Continuity Condition) $f \in C^2(\mathbb{R}^n)$ satisfies the local gradient Lipschitz continuity condition at $x_0 \in \mathbb{R}^n$ with radius $\delta > 0$ if there exists a constant $L_{(x_0,\delta)} > 0$ such that

$$\|\nabla f(x) - \nabla f(y)\|_2 \leq L_{(x_0,\delta)} \|x - y\|_2 \qquad \forall x, y \in V(x_0, \delta). \tag{5.4}$$

We next review the concepts of *stationary point*, *local minimizer*, and *strict saddle point*, which are important in (non-convex) optimization.

Definition 5.4 (Stationary Point) $x^* \in \mathbb{R}^n$ is a *stationary point* of $f \in C^2(\mathbb{R}^n)$ if $\nabla f(x^*) = 0$.

Definition 5.5 (Local Minimizer) $x^* \in \mathbb{R}^n$ is a local minimum of f if there is a neighborhood U around x^* such that for all $x \in U$, $f(x^*) < f(x)$.

A stationary point can be a local minimizer, a saddle point, or a maximizer. It is a standard fact that if a stationary point $x^\star \in \mathbb{R}^n$ is a local minimizer of $f \in C^2(\mathbb{R}^n)$, then $\nabla^2 f(x^\star)$ is positive semidefinite; on the other hand, if $x^* \in \mathbb{R}^n$ is a stationary point of $f \in C^2(\mathbb{R}^n)$ and $\nabla^2 f(x^\star)$ is positive definite, then x^* is also a local minimizer of f. It should also be noted that the stationary point $x^\star$ in the second case is isolated.

The following definition concerns "strict" saddle points, which was also analyzed in [GHJY15].

Definition 5.6 (Strict Saddle Points) $x^* \in \mathbb{R}^n$ is a *strict saddle*[1] of $f \in C^2(\mathbb{R}^n)$ if x^* is a stationary point of f and furthermore $\lambda_{\min}\left(\nabla^2 f(x^*)\right) < 0$.

We denote the set of all strict saddle points by $\mathcal{X}$. By definition, a strict saddle point must have an escaping direction so that the eigenvalue of the Hessian along that direction is strictly negative. For many non-convex problems studied in machine learning, all saddle points are strict.

We next review additional concepts in multivariate analysis and differential geometry/topology that will be used in our analysis.

Definition 5.7 (Gradient Map and Its Jacobian) For any $f \in C^2(\mathbb{R}^n)$, the gradient map $g : \mathbb{R}^n \to \mathbb{R}^n$ with step size h is defined as

$$g(x) = x - h\nabla f(x). \tag{5.5}$$

[1]For the purposes, strict saddle points include local maximizers.

The Jacobian $Dg : \mathbb{R}^n \to \mathbb{R}^{n\times n}$ of the gradient map g is defined as

$$Dg(x) = \begin{pmatrix} \frac{\partial g_1}{\partial x_1}(x) & \cdots & \frac{\partial g_1}{\partial x_n}(x) \\ \cdots & \cdots & \cdots \\ \frac{\partial g_n}{\partial x_1}(x) & \cdots & \frac{\partial g_n}{\partial x_n}(x) \end{pmatrix}, \tag{5.6}$$

or equivalently, $Dg = I - h\nabla^2 f$.

We write $a_n \lesssim b_n$ if there exists an absolute constant $C > 0$ such that, for sufficiently large n, $|a_n| \leq C|b_n|$. Similarly, $a_n \gtrsim b_n$ if $b_n \lesssim a_n$ and $a_n \asymp b_n$ if both $a_n \lesssim b_n$ and $b_n \lesssim a_n$ are true. We write $a_n \ll b_n$ if for a sufficiently small constant $c > 0$ and sufficiently large n, $|a_n| \leq c|b_n|$. For any integer M, $[M]$ denotes the finite set $\{1, 2, \cdots, M\}$.

Definition 5.8 (Local Diffeomorphism) Let M and N be two differentiable manifolds. A map $f : M \to N$ is a *local diffeomorphism* if for each point x in M, there exists an open set U containing x such that $f(U)$ is open in N and $f|_U : U \to f(U)$ is a diffeomorphism.

Definition 5.9 (Compact Set) $S \subseteq \mathbb{R}^n$ is *compact* if every open cover of S has a finite sub-cover.

Definition 5.10 (Sublevel Set) The α-sublevel set of $f : \mathbb{R}^n \to \mathbb{R}$ is defined as

$$C_\alpha = \left\{x \in \mathbb{R}^n \mid f(x) \leq \alpha\right\}.$$

5.1 Concluding Remarks

In this chapter, we have defined the necessary notations that would be used in the remaining part of this book. We have also identified and reviewed the important definitions that will be used later in our analysis. The following chapter discusses the related work on the geometry of the non-convex programs.

Chapter 6
Related Work on Geometry of Non-Convex Programs

Over the past few years, there have been increasing interest in understanding the geometry of non-convex programs that naturally arise from machine learning problems. It is particularly interesting to study additional properties of the considered non-convex objective such that popular optimization methods (such as gradient descent) escape saddle points and converge to a local minimum. The strict saddle property (Definition 5.6) is one such property, which was also shown to hold in a broad range of applications.

Many existing works leveraged Hessian information to circumvent saddle points. This includes a modified Newton's method [MS79], the modified Cholesky's method [GM74], the cubic-regularization method [NP06], and trust region methods [CRS14]. The major drawback of such second-order methods is the requirement of access to the full Hessian, which could be computationally expensive, as the per-iteration computational complexity scales quadratically or even cubically in the problem dimension, unsuitable for optimization of high-dimensional functions. Some recent works [CDHS16, AAB$^+$17, CD16] showed that the requirement of full Hessian can be relaxed to Hessian-vector products, which can be computed efficiently in certain machine learning applications. Several works [LY17, RZS$^+$17, RW17] also presented algorithms that combine first-order methods with faster eigenvector algorithms to obtain lower per-iteration complexity.

Another line of works focuses on noise-injected gradient methods whose per-iteration computational complexity scales linearly in the problem dimension. Earlier work has shown that the first-order method with unbiased noise with sufficiently large variance can escape strict saddle points [Pem90]. Ge et al. [GHJY15] gave quantitative rates on the convergence. Recently, more refined algorithms and analyses [JGN$^+$17, JNJ17] have been proposed to improve the convergence rate of such algorithms. Nevertheless, gradient methods with *deliberately injected* noise are almost never used in practical applications, limiting the applicability of the above-mentioned analysis.

B. Shi, S. S. Iyengar, *Mathematical Theories of Machine Learning - Theory and Applications*, https://doi.org/10.1007/978-3-030-17076-9_6

Empirically, [SQW16] observed that gradient descent with 100 random initializations for the phase retrieval problem always converges to a local minimizer. Theoretically, the most important existing result is due to [LSJR16], who showed that gradient descent with fixed step size and any reasonable random initialization always escapes isolated strict saddle points. Panageas and Piliouras [PP16] later relaxed the requirement that strict saddle points are isolated. O'Neill and Wright [OW17] extended the analysis to accelerated gradient descent and [LPP$^+$17] generalized the result to a broader range of first-order methods, including proximal gradient descent and coordinate descent. However these works all require the step size to be significantly smaller than the inverse of Lipschitz constant of gradients, which has factor of 2 gap from results in the convex setting and do not allow the step size to vary across iterations. Our work resolves both problems.

The history of the use of the gradient method for convex optimization dates back to the time of Euler and Lagrange. However, its relatively cheaper means of performing the calculations for the first-order information makes it a method still in active use in machine learning and non-convex optimization, such as the recent work in [GHJY15, AG16, LSJR16, HMR16]. The natural speedup algorithms are the momentum method first proposed in [Pol64] and Nesterov's accelerated gradient method first proposed in [Nes83] with an improved version [NN88]. An acceleration algorithm in similar lines to Nesterov's accelerated gradient method, called FISTA, is designed to solve composition problems as depicted in [BT09]. A related comprehensive work is proposed in [B$^+$15].

The original momentum method, named as Polyak heavy ball method, is from the view of ODE in [Pol64], which contains extremely rich physical intuitive ideas and mathematical theory. An extremely important work in application on machine learning is the backpropagation learning with momentum [RHW$^+$88]. Based on the thought of ODE, a lot of understanding and application on the momentum method and Nesterov's accelerated gradient methods have been proposed. In [SMDH13], a well-designed random initialization with momentum parameter algorithm is proposed to train both DNNs and RNNs. A breakthrough deep insight from ODE to understanding the intuition behind the Nesterov scheme is proposed in [SBC14]. The understanding for momentum method based on the variation perspective is proposed in [WWJ16], and the understanding from Lyapunov analysis is proposed in [WRJ16]. From the stability theorem of ODE, the gradient method always converges to local minima in the sense of almost everywhere is proposed in [LSJR16]. Analyzing and designing iterative optimization algorithms built on integral quadratic constraints from robust control theory is proposed in [LRP16].

Actually the "high momentum" phenomenon was initially observed in [OC15] for a restarting adaptive accelerating algorithm. A restarting scheme has been proposed in the work by [SBC14]. However, both these above-mentioned works use restarting scheme for an auxiliary tool to accelerate the algorithm based on friction. Utilizing the concept of phase space in mechanics, we can observe that the kinetic energy, or velocity, is controllable and can be molded to being an utilizable parameter that would aid in finding the local minima. Without the use of the term defining friction, we still would be able to find the local minima by using just the velocity parameter. For this advantage, this algorithm is touted to be very simplistic to practice. We can generalize this to the non-convex optimization in the detection of the local minima along the trajectory of the particle.

Sparse subspace clustering was proposed by [EV13] as an effective method for subspace clustering. Soltanolkotabi and Candes [SC12] initiated the study of theoretical properties of sparse subspace clustering, which was later extended to noisy data [SEC14, WX16], dimensionality-reduced data [WWS15a, HTB17, TG17], and data consisting of sensitive private information [WWS15b]. Yang et al. [YRV15] considered some heuristics for subspace clustering with missing entries, and [TV18] considered a PZF-SSC approach and proved success conditions with $\rho = \Omega(1)$. Park et al. [PCS14], Heckel and Bölcskei [HB15], Liu et al. [LLY$^+$13], Tsakiris and Vidal [TV17] proposed alternative approaches for subspace clustering. Some earlier references include k-plane [BM00], q-flat [Tse00], ALC [MDHW07], LSA [YP06], and GPCA [VMS05].

It is an important task to reveal the casual dependencies between historical and current observations in MTS analysis. Bayesian Network [JYG$^+$03, Mur02] and Granger causality [ALA07, ZF09] are two main frameworks for inference of temporal dependency. Comparing with Bayesian Network, Granger causality is more straightforward, robust, and extendable [ZF09].

The online adaptive linear regression cannot only provide insights about the concerned time series with time evolution, but also is essential for parameter estimation, prediction, cross prediction, i.e., predicting how the variables depend on the other variables, and so on. In this paper, we are interested in uncovering the dynamical relationship, which characterizes the time-specific dependencies between variables.

Temporal data sets are a collection of data items associated with time stamps. It can be divided into two categories, i.e., time-series data and event data. Just as time-series data, the essential difference between univariate and multivariate is the dependent relationship among variables. Our work focuses on multivariate time-series data sets.

6.1 Multivariate Time-Series Data Sets

Originally, Granger causality is designed for a pair of time series. The appearance of pioneering work of combining the notion of Granger causality with graphical model [Eic06] leads to the emergence of causal relationship analysis among MTS data. Two typical techniques, statistical significance test and Lasso-Granger [ALA07], are developed to inference the Granger causality among MTS. Lasso-Granger gains more popularity due to its robust performance even in high dimensions [BL12]. However, Lasso-Granger suffers from instability and failure of group variable selection because of the high sensitivity of L_1 norm. To address this challenging, our method adopts elastic-net regularizer [ZH05] which is stable since it encourages a group variable selection (group effect) where strongly correlated predictors tend to be zero or non-zero simultaneously.

Our proposed method utilizes the advantages of Lasso-Granger, but conducts the inference from the Bayesian perspective in a sequential online mode, borrowing the ideas of Bayesian Lasso [TPGC]. However, most of these methods assume a constant dependency structure among the time series.

Regression model has evolved to be one of the principal approaches for Granger causality and most of the existing methods assume a constant dependency structure, i.e., a constant causal relationship between time series. One is the Bayesian network inference approach [Hec98, Mur12, JYG+03], while the other approach is the Granger causality [Gra69, Gra80, ALA07]. An extensive comparison study between these two types of frameworks is presented in [ZF09]. In order to overcome these difficulties, the time-varying Lasso-Granger model based on an online manner is proposed [ZWW+16]. However, the Lasso regularizer has its own limit, which cannot capture the natural group information between variables and extremely instability. Advances in regularization theory have led to a series of extensions to the original Lasso algorithm, such as elastic-net [ZH05].

Related to the elastic-net regularizer, the offline algorithms first are proposed for capturing the relationship in [LLNM+09]. The elastic net encourages a grouping effect, where strongly correlated predictors tend to be in or out of the model together. Real-world data and a simulation study show that elastic net often outperforms lasso while enjoying a similar sparse representation [ZH05].

Based on the online algorithm of particle learning for high-dimensional linear regression firstly borrowed by [ZWML16], we introduce the elastic-net regularizer to the online algorithm and investigate the group effects among variables.

Particle learning [CJLP10] is a powerful tool to provide an online inference strategy for Bayesian models. It belongs to the sequential Monte Carlo (SMC) methods consisting of a set of Monte Carlo methodologies to solve the filtering problem [DGA00]. Particle learning provides state filtering, sequential parameter learning, and smoothing in a general class of state space models [CJLP10]. The central idea behind particle learning is the creation of a particle algorithm that directly samples from the particle approximation to the joint posterior distribution of states and conditional sufficient statistics for fixed parameters in a fully adapted resample-propagate framework.

We borrow the idea of particle learning for both latent state inference and parameter learning.

Zeng et al. [ZWW+16] handle and simulate the time-varying multivariate time series. In a general class of state space models, particle learning provides state filtering, sequential parameter learning, and smoothing. Particle learning is for approximating the sequence of filtering and smoothing distributions in light of parameter uncertainty. The central idea behind particle learning is the creation of a particle algorithm that directly samples from the particle approximation to the joint posterior distribution of states and conditional sufficient statistics for fixed parameters in a fully adapted resample-propagate framework.

Multivariate time series is a very important tool to try to explain this phenomenon. The spatial–temporal causal modeling is first proposed in [LLNM+09]. Based on the graph inference, the relational graph model is proposed in [LNMLL10] and the varying graph model in [CLLC10]. Also, the sparse-GEV model of latent state on extreme value is proposed in [LBL12]. However, all of the models are static, which cannot capture the dynamical information.

6.2 Particle Learning

To capture dynamic information in state space modeling, researchers have incorporated additional information through sequential parameter learning. Limitations in applying these methods must be considered, as many computational challenges result from these applications. State filtering, sequential parameter learning, and smoothing in a general class of state space models have recently been applied [CJLP10]. Of these, particle learning has become one of the most popular sequential learning methods.

The central idea behind particle learning is the creation of a particle algorithm that directly samples from the particle approximation to the joint posterior distribution of states and conditional sufficient statistics for fixed parameters in a fully adapted resample-propagate framework. The idea of particle learning for both latent state inference and parameter learning is firstly borrowed by [ZWML16]. Here, we continue to make use of the idea on our elastic-net regularizer.

6.3 Application to Climate Change

Climate change poses many critical social issues in the new century. Uncovering the dependency relationship between the various climate observations and forcing factors is an important challenge. Multivariate time series is a very important tool to try to explain this phenomenon. The spatial–temporal causal modeling is first proposed in [LLNM+09]. Based on the graph inference, the relational graph model is proposed in [LNMLL10] and the varying graph model in [CLLC10]. Also, the sparse-GEV model of latent state on extreme value is proposed in [LBL12]. However, all of the models are static, which cannot capture the dynamical information. Here we proposed an online time-varying spatial–temporal causal model to simulate and interpret the phenomena of air temperature in Florida.

6.4 Concluding Remarks

In this chapter, we have reviewed and discussed the various ideas and concepts related to the geometry of the non-convex programs. We have also described the concepts of particle learning and an introduction to one of the applications "climate change" with a case study of Florida.

In the next part of the book, we discuss the mathematical framework for machine learning giving emphasis to the theoretical aspects.

Part II
Mathematical Framework for Machine Learning: Theoretical Part

Chapter 7
Gradient Descent Converges to Minimizers: Optimal and Adaptive Step-Size Rules

7.1 Introduction

As mentioned in Chap. 3, gradient descent (GD) and its variants provide the core optimization methodology in machine learning problems. Given a C^1 or C^2 function $f : \mathbb{R}^n \to \mathbb{R}$ with unconstrained variable $x \in \mathbb{R}^n$, GD uses the following update rule:

$$x_{t+1} = x_t - h_t \nabla f(x_t) \tag{7.1}$$

where h_t are step size, which may be either fixed or vary across iterations. When f is convex, $h_t < \frac{2}{L}$ is a necessary and sufficient condition to guarantee the (worst-case) convergence of GD, where L is the Lipschitz constant of the gradient of the function f. On the other hand, there is far less understanding of GD for non-convex problems. For general smooth non-convex problems, GD is only known to converge to a stationary point (i.e., a point with zero gradient) [Nes13].

Machine learning tasks often require finding a local minimizer instead of just a stationary point, which can also be a saddle point or a maximizer. In recent years, there has been an increasing focus on geometric conditions under which GD escapes saddle points and converges to a local minimizer. More specifically, if the objective function satisfies (1) all saddle point are strict and (2) all local minima are global minima, then GD finds a global optimal solution. These two properties hold for a

Part of this chapter is in the paper titled "gradient decent converges to minimizers: optimal and adaptive step size rules" by Bin Shi et al. (2018) presently under review for publication in INFORMS, Journal on Optimization.

B. Shi, S. S. Iyengar, *Mathematical Theories of Machine Learning - Theory and Applications*, https://doi.org/10.1007/978-3-030-17076-9_7

wide range of machine learning problems, such as matrix factorization [LWL$^+$16], matrix completion [GLM16, GJZ17], matrix sensing [BNS16, PKCS17], tensor decomposition [GHJY15], dictionary learning [SQW17], and phase retrieval [SQW16].

Recent works showed when the objective function has the strict saddle property, then GD converges to a minimizer provided the initialization is randomized and the step sizes are fixed and smaller than $1/L$ [LSJR16, PP16]. While this was the first result establishing convergence of GD, there are still gaps towards fully understanding GD for strict saddle problems. In particular, as mentioned in Chap. 4, the following two questions are still open regarding the convergence of GD for non-convex problems:

Question 1: Maximum Allowable Fixed Step Size Recall that for convex optimization by gradient decent with fixed step-size rule $h_t \equiv h$, $h < 2/L$ is both a necessary and a sufficient condition for the convergence of GD. However, for non-convex optimization existing works all required the (fixed) step size to be smaller than $1/L$. Because larger step sizes lead to faster convergence, a natural question is to identify the maximum allowable step size such that GD escapes saddle points. The main technical difficulty to analyze larger step size is that the gradient map

$$g(x) = x - h\nabla f(x)$$

may *not* be a diffeomorphism when $h \geq 1/L$. Thus, techniques used in [LSJR16, PP16] are no longer sufficient.

In this paper, we take a finer look at the dynamics of GD. Our main observation is that the GD procedure escapes strict saddle points under much weaker conditions than g being a diffeomorphism everywhere. In particular, the probability of GD with random initialization converging to a strict saddle point is 0 provided that

$$g(x_t) = x_t - h_t\nabla f(x_t)$$

is a *local* diffeomorphism at every x_t. We further show that

$$\lambda\left(\{h \in [1/L, 2/L) : \exists t, g(x_t) \text{ is not a local diffeomorphism}\}\right) = 0,$$

where $\lambda(\cdot)$ is the standard Lebesgue measure on $\mathbb{R}$, meaning that for almost every fixed step size choice in $[1/L, 2/L)$, $g(x_t)$ is a local diffeomorphism for every t. Therefore, if a step size h is chosen uniformly at random from $\left(\frac{2}{L} - \epsilon, \frac{2}{L}\right)$ for any $\epsilon > 0$, GD escapes all strict saddle points and converges to a local minimum. See Sect. 7.3 for the precise statement and Sect. 7.5 for the proof.

Question 2: Analysis of Adaptive Step Sizes Another open question we consider is to analyze the convergence of GD for non-convex objectives when the step sizes $\{h_t\}$ vary as t evolves. In convex optimization, adaptive step-size rules such as exact or backtracking line search [Nes13] are commonly used in practice to improve convergence, and convergence of GD is guaranteed provided that the adaptively tuned step sizes do not exceed twice the inverse of local gradient Lipschitz constant. On the other hand, in non-convex optimization, whether gradient descent with varying step sizes can escape all strict saddle points is unknown.

Existing techniques [LSJR16, PP16, LPP$^+$17, OW17] cannot solve this question because they relied on the classical stable manifold theorem [Shu13], which requires a fixed gradient map, whereas when the step sizes vary, the gradient maps also change across iterations. To deal with this issue, we adopt the powerful Hartman product map theorem [Har71], which gives a finer characterization of local behavior of GD and allows the gradient map to change at every iteration. Based on Hartman product map theorem, we show that as long as the step size at each iteration is proportional to the inverse of the *local* gradient Lipschitz constant, GD still escapes all strict saddle points. To the best of our knowledge, this is the first result establishing convergence to local minima for non-convex gradient descent with varying step sizes.

7.1.1 Related Works

Over the past few years, there have been increasing interest in understanding the geometry of non-convex programs that naturally arise from machine learning problems. It is particularly interesting to study additional properties of the considered non-convex objective such that popular optimization methods (such as gradient descent) escape saddle points and converge to a local minimum. The strict saddle property (Definition 7.6) is one such property, which was also shown to hold in a broad range of applications.

Many existing works leveraged Hessian information in order to circumvent saddle points. This includes a modified Newton's method [MS79], the modified Cholesky's method [GM74], the cubic-regularization method [NP06], and trust region methods [CRS14]. The major drawback of such second-order methods is the requirement of access to the full Hessian, which could be computationally expensive, as the per-iteration computational complexity scales quadratically or even cubically in the problem dimension, unsuitable for optimization of high-dimensional functions. Some recent works [CDHS16, AAB$^+$17, CD16] showed that the requirement of full Hessian can be relaxed to Hessian-vector products, which can be computed efficiently in certain machine learning applications. Several works [LY17, RZS$^+$17, RW17] also presented algorithms that combine first-order methods with faster eigenvector algorithms to obtain lower per-iteration complexity.

Another line of works focuses on noise-injected gradient methods whose per-iteration computational complexity scales linearly in the problem dimension. Earlier work have shown that first-order method with unbiased noise with sufficiently large variance can escape strict saddle points [Pem90]. [GHJY15] gave quantitative rates on the convergence. Recently, more refined algorithms and analyses [JGN$^+$17, JNJ17] have been proposed to improve the convergence rate of such algorithms.

Nevertheless, gradient methods with *deliberately injected* noise are almost never used in practical applications, limiting the applicability of the above-mentioned analysis.

Empirically, [SQW16] observed that gradient descent with 100 random initializations for the phase retrieval problem always converges to a local minimizer. Theoretically, the most important existing result is due to [LSJR16], who showed that gradient descent with fixed step size and any reasonable random initialization always escapes isolated strict saddle points. [PP16] later relaxed the requirement that strict saddle points are isolated. Other authors, [OW17] extended the analysis to accelerated gradient descent and [LPP+17] generalized the result to a broader range of first-order methods, including proximal gradient descent and coordinate descent. However these works all require the step size to be significantly smaller than the inverse of Lipschitz constant of gradients, which has factor of 2 gap from results in the convex setting and do not allow the step size to vary across iterations. Our results resolve both the aforementioned problems.

7.2 Notations and Preliminaries

For the sake of completeness, we present necessary notations and review important definitions some of which defined earlier in Chap. 5 and will be used later in our analysis. Let $C^2(\mathbb{R}^n)$ be the vector space of real-valued twice-continuously differentiable functions. Let ∇ be the gradient operator and ∇^2 be the Hessian operator. Let $\|\cdot\|_2$ be the Euclidean norm in $\mathbb{R}^n$. Let μ be the Lebesgue measure in $\mathbb{R}^n$.

Definition 7.1 (Global Gradient Lipschitz Continuity Condition) $f \in C^2(\mathbb{R}^n)$ satisfies the global gradient Lipschitz continuity condition if there exists a constant $L > 0$ such that

$$\|\nabla f(x_1) - \nabla f(x_2)\|_2 \le L \|x_1 - x_2\|_2 \qquad \forall x_1, x_2 \in \mathbb{R}^n. \tag{7.2}$$

Definition 7.2 (Global Hessian Lipschitz Continuity Condition) $f \in C^2(\mathbb{R}^n)$ satisfies the global Hessian Lipschitz continuity condition if there exists a constant $K > 0$ such that

$$\left\|\nabla^2 f(x_1) - \nabla^2 f(x_2)\right\|_2 \le K \|x_1 - x_2\|_2 \qquad \forall x_1, x_2 \in \mathbb{R}^n. \tag{7.3}$$

Intuitively, a twice-continuously differentiable function $f \in C^2(\mathbb{R}^n)$ satisfies the global gradient and Hessian Lipschitz continuity condition if its gradients and Hessians do not change dramatically for any two points in $\mathbb{R}^n$. However, the global Lipschitz constant L for many objective functions that arise in machine learning applications (e.g., $f(x) = x^4$) may be large or even non-existent. To handle such cases, one can use a finer definition of gradient continuity that characterizes the *local* behavior of gradients, especially for non-convex functions. This definition is adopted in many subjects of mathematics, such as in dynamical systems research.

Let $\delta > 0$ be some fixed constant. For every $x_0 \in \mathbb{R}^n$, its δ-closed neighborhood is defined as

$$V(x_0, \delta) = \left\{ x \in \mathbb{R}^n \middle| \|x - x_0\|_2 < \delta \right\}. \tag{7.4}$$

Definition 7.3 (Local Gradient Lipschitz Continuity Condition) The function $f \in C^2(\mathbb{R}^n)$ satisfies the local gradient Lipschitz continuity condition at $x_0 \in \mathbb{R}^n$ with radius $\delta > 0$ if there exists a constant $L_{(x_0,\delta)} > 0$ such that

$$\|\nabla f(x) - \nabla f(y)\|_2 \leq L_{(x_0,\delta)} \|x - y\|_2 \qquad \forall x, y \in V(x_0, \delta). \tag{7.5}$$

We next review the concepts of *stationary point*, *local minimizer*, and *strict saddle point*, which are important in (non-convex) optimization.

Definition 7.4 (Stationary Point) $x^* \in \mathbb{R}^n$ is a *stationary point* of $f \in C^2(\mathbb{R}^n)$ if $\nabla f(x^*) = 0$.

Definition 7.5 (Local Minimizer) $x^* \in \mathbb{R}^n$ is a local minimum of f if there is a neighborhood U around x^* such that for all $x \in U$, $f(x^*) < f(x)$.

A stationary point can be a local minimizer, a saddle point, or a maximizer. It is a standard fact that if a stationary point $x^\star \in \mathbb{R}^n$ is a local minimizer of $f \in C^2(\mathbb{R}^n)$, then $\nabla^2 f(x^\star)$ is positive semidefinite; on the other hand, if $x^* \in \mathbb{R}^n$ is a stationary point of $f \in C^2(\mathbb{R}^n)$ and $\nabla^2 f(x^\star)$ is positive definite, then x^* is also a local minimizer of f. It should also be noted that the stationary point $x^\star$ in the second case is isolated.

The following definition concerns "strict" saddle points, which was also analyzed in [GHJY15].

Definition 7.6 (Strict Saddle Points) $x^* \in \mathbb{R}^n$ is a *strict saddle*[1] of $f \in C^2(\mathbb{R}^n)$ if x^* is a stationary point of f and furthermore $\lambda_{\min}\left(\nabla^2 f(x^*)\right) < 0$.

We denote the set of all strict saddle points by $\mathcal{X}$. By definition, a strict saddle point must have an escaping direction so that the eigenvalue of the Hessian along that direction is strictly negative. For many non-convex problems studied in machine learning, all saddle points are strict.

We next review additional concepts in multivariate analysis and differential geometry/topology that will be used in our analysis.

Definition 7.7 (Gradient Map and Its Jacobian) For any $f \in C^2(\mathbb{R}^n)$, the gradient map $g : \mathbb{R}^n \to \mathbb{R}^n$ with step size h is defined as

$$g(x) = x - h\nabla f(x). \tag{7.6}$$

[1]For the purpose of this paper, strict saddle points include local maximizers.

The Jacobian $Dg : \mathbb{R}^n \to \mathbb{R}^{n\times n}$ of the gradient map g is defined as

$$Dg(x) = \begin{pmatrix} \frac{\partial g_1}{\partial x_1}(x) & \cdots & \frac{\partial g_1}{\partial x_n}(x) \\ \cdots & \cdots & \cdots \\ \frac{\partial g_n}{\partial x_1}(x) & \cdots & \frac{\partial g_n}{\partial x_n}(x) \end{pmatrix}, \tag{7.7}$$

or equivalently, $Dg = I - h\nabla^2 f$.

Definition 7.8 (Local Diffeomorphism) Let M and N be two differentiable manifolds. A map $f : M \to N$ is a *local diffeomorphism* if for each point x in M, there exists an open set U containing x such that $f(U)$ is open in N and $f|_U : U \to f(U)$ is a diffeomorphism.

Definition 7.9 (Compact Set) $S \subseteq \mathbb{R}^n$ is *compact* if every open cover of S has a finite sub-cover.

Definition 7.10 (Sublevel Set) The α-sublevel set of $f : \mathbb{R}^n \to \mathbb{R}$ is defined as

$$C_\alpha = \left\{x \in \mathbb{R}^n \mid f(x) \le \alpha\right\}.$$

7.3 Maximum Allowable Step Size

We first consider gradient descent with a fixed step size. The following theorem provides a sufficient condition for escaping all strict saddle points.

Theorem 7.1 *Suppose $f \in C^2(\mathbb{R}^n)$ satisfies the global gradient Lipschitz condition (Definition 7.1) with constant $L > 0$. Then there exists a zero-measure set $U \subset \left[\frac{1}{L}, \frac{2}{L}\right)$ such that if $h \in \left(0, \frac{2}{L}\right) \setminus U$ and $x_0 \in \mathbb{R}^n$ is randomly initialized with respect to an absolute continuous measure over $\mathbb{R}^n$, then*

$$\mathbf{Pr}\left(\lim_k x_k \in \mathcal{X}\right) = 0,$$

where $\mathcal{X}$ denotes the set of all strict saddle points of f.

The complete proof of Theorem 7.1 is given in Sect. 7.5. Here we give a high-level sketch of our proof. Similar to [LSJR16], our proof relies on the seminal stable manifold Theorem [Shu13]. For a fixed saddle point x^*, the stable manifold theorem asserts that locally, all points that converge to x^* lie in a manifold $W_{loc}^{cs}(x^*)$. Further, $W_{loc}^{cs}(x^*)$ has dimension at most $n-1$, thus $\mu\left(W_{loc}^{cs}(x^*)\right) = 0$. By Lindelöf's lemma (Lemma 7.6), we can show that the union of these manifolds, $W_{loc}^{cs} = \bigcup_{x^*\in\mathcal{X}} W_{loc}^{cs}(x^*)$, also has Lebesgue measure 0. Next, we analyze what initialization points converge to W_{loc}^{cs}. Using the notion of the inverse gradient map, we can show that the initialization points that converge W_{loc}^{cs} belong to the set

$$\bigcup_{i=0}^{\infty} g^{-i}(W_{loc}^{cs}).$$

Thus, we only need to upper bound the Lebesgue measure of this set. If g is a local diffeomorphism, then by Lemma 7.2, we have $\mu\left(\bigcup_{i=0}^{\infty} g^{-i}(W_{loc}^{cs})\right) \leq \sum_{i=0}^{\infty} \mu\left(g^{-i}\left(W_{loc}^{cs}\right)\right) = 0$. Therefore, we only need to show g is a local diffeomorphism. Existing works require $\eta \leq 1/L$ to ensure g is a *global* diffeomorphism, whereas a *local* diffeomorphism is already sufficient. Our main observation is that for h in $(1/L, 2/L)$, there is only a zero-measure set U such that g with respect to $h \in U$ is *not* a local diffeomorphism at some x_t. In other words, for almost every step size $h \in (1/L, 2/L)$, g is a local diffeomorphism at x_t for every t.

Theorem 7.1 shows that the step sizes in $[1/L, 2/L)$ that potentially leads to GD convergence towards a strict saddle point have measure zero. Comparing to recent results on gradient descent by [LSJR16, LPP$^+$17, PP16], our theorem allows a maximum (fixed) step size of $2/L$ instead of $1/L$.

7.3.1 Consequences of Theorem 7.1

A direct corollary of Theorem 7.1 is that GD (with fixed step sizes $< 2/L$) can only converge to minimizers when the limit $\lim_k x_k$ exists.

Corollary 7.1 (GD Converges to Minimizers) *Under the conditions in Theorem 7.1 and the additional assumption that all saddle points of f are strict, if* $\lim_k x_k$ *exists then with probability 1* $\lim_k x_k$ *is a local minimizer of f.*

We now discuss when $\lim_k x_k$ exists. The following lemma gives a sufficient condition on its existence.

Lemma 7.1 *Suppose $f \in C^2(\mathbb{R}^n)$ has global gradient Lipschitz constant L and owns compact sublevel sets. Further assume f only contains isolated stationary points. If $0 < h < 2/L$,* $\lim_k x_k$ *converges to a stationary point of f for any initialization x_0.*

Theorem 7.1 and Lemma 7.1 together imply Corollary 7.1, which asserts that if the objective function has compact sublevel sets and the fixed step size h is smaller than $2/L$, GD converges to a minimizer. This result generalizes [LSJR16, PP16] where the fixed step sizes of GD cannot exceed $1/L$.

7.3.2 Optimality of Theorem 7.1

A natural question is whether the condition $h < 2/L$ in Theorem 7.1 can be further improved. The following proposition gives a negative answer, showing that GD with fixed step sizes $h \geq 2/L$ *diverges* on worst-case objective function f with probability 1. This shows that $h < 2/L$ is the optimal fixed step-size rule one can hope for with which GD converges to a local minimum almost surely.

Proposition 7.1 *There exists $f \in C^2(\mathbb{R}^n)$ with global gradient Lipschitz constant $L > 0$, compact sublevel sets, and only isolated stationary points such that if $h \geq 2/L$ and x_0 is randomly initialized with respect to an absolutely continuous density on $\mathbb{R}^n$, then $\lim_k x_k$ does not exist with probability 1.*

The proof of the proposition is simple by considering a quadratic function $f \in C^2(\mathbb{R}^n)$ that serves as a counter-example of GD with fixed step sizes larger than or equal to $h/2$. A complete proof of Proposition 7.1 is given in the appendix.

7.4 Adaptive Step-Size Rules

In many machine learning applications, the global gradient Lipschitz constant L of the objective function f may be very large, but at most points the *local* gradient Lipschitz constant could be much smaller. It is thus desirable to consider varying step-size rules that select step sizes h_t *adaptively* corresponding to the local gradient Lipschitz constant of f at x_t. When the objective function f is convex, the convergence of gradient descent with varying step sizes is well-understood [Nes13]. However, when f is non-convex, whether GD with varying step sizes can escape strict saddle points is still unknown. Existing works [LSJR16, LPP$^+$17, PP16] all require the step sizes to be fixed. Our following result closes this gap, showing that GD escapes strict saddle points if the step sizes chosen at each point x_t are proportional to the local gradient Lipschitz constant $L_{x_t,\delta}$.

Theorem 7.2 *Suppose $f \in C^2(\mathbb{R}^n)$ satisfies the global Hessian Lipschitz continuity condition (Definition 7.2) with parameter K and for every $x^* \in \mathcal{X}$, $\nabla^2 f(x^*)$ is non-singular. Fix $\epsilon_0 \in (0, 1)$ and define $r = \max_{x^* \in \mathcal{X}} K^{-1}\epsilon_0 \|\nabla^2 f(x^*)\|_2$. Then there exists $U \subset \mathbb{R}^+$ with $\mu(U) = 0$ such that if the step size at the tth iteration satisfies $h_t \in \left[\frac{\epsilon_0}{L_{x_t,r}}, \frac{2-\epsilon_0}{L_{x_t,r}}\right] \setminus U$ for all $t = 0, 1, \ldots$ and x_0 is randomly initialized with respect to an absolutely continuous density on $\mathbb{R}^n$, then*

$$\mathbf{Pr}\left(\lim_t x_t \in \mathcal{X}\right) = 0.$$

Theorem 7.2 shows that even though the step sizes vary across iterations, GD still escapes all strict saddle points provided that all step sizes are proportional to their *local* smoothness. To the best of our knowledge, this is the first result showing that GD with varying step size escapes all strict saddle points. Theorem 7.2 requires $h_t \in \left[\frac{\epsilon_0}{L_{x_t,\delta}}, \frac{2-\epsilon_0}{L_{x_t,\delta}}\right]$, which are the desired local step size.

The proof of Theorem 7.2 follows a similar path as that of Theorem 7.1. We first locally characterize Lebesgue measure of the set of points that converge to saddle points and then use Lemma 7.2 to relate this set to the initialization. The main technical difficulty is the inapplicability of the stable manifold theorem in this setting, as the gradient maps g are no longer *fixed* and change across iterations. Instead of using the stable manifold theorem, we adopt the more general Hartman's product map theorem [Har82] which gives a finer characterization of the local behavior of a series of gradient maps around a saddle point.

Different from Theorems 7.1 and 7.2 has two additional assumptions. First, we require that the Hessian matrices $\nabla^2 f(x^*)$ at each saddle point x^* are non-singular (i.e., no zero eigenvalues). This is a technical regularity condition for using Hartman's product map theorem. To remove this assumption, we need to generalize Hartman's product map theorem which is a challenging problem in dynamical systems. Second, we require that Hessian matrices $\nabla^2 f(x)$ satisfy a global Lipschitz continuity condition (Definition 7.2). This is because the Hartman's product map theorem requires the step size to be proportional to the gradient Lipschitz constant in a neighborhood of each saddle point and the radius of the neighborhood needs to be carefully quantified. Under the Hessian Lipschitz continuity assumption, we can give an upper bound on this radius which is sufficient for applying Hartman's product map theorem. It is possible to give finer upper bounds on this radius based on other quantitative continuity assumptions on the Hessian. The complete proof of Theorem 7.2 is given in Sect. 7.6.

7.5 Proof of Theorem 7.1

To prove Theorem 7.1, similar to [LSJR16], we rely on the following seminal stable manifold theorem from dynamical systems research.

Theorem 7.3 (Theorem III. 7, p. 65, [Shu13]) *Let* 0 *be a fixed point for a* C^r *local diffeomorphism* $f : U \to \mathbb{R}^n$, *where* U *is a neighborhood of zero in* $\mathbb{R}^n$ *and* $1 \le r < \infty$. *Let* $E^s \bigoplus E^c \bigoplus E^u$ *be the invariant splitting of* $\mathbb{R}^n$ *into the generalized eigenspaces of* $Df(0)$ *corresponding to eigenvalues of absolute value less than one, equal to one, and greater than one. To the* $Df(0)$ *invariant subspaces* $E^s \bigoplus E^c$, E^c *there is associated a local* f *invariant* C^r *embedded disc* W^{cs}_{loc} *tangent to the linear subspace at* 0 *and a ball* B *around zero in an adapted norm such that*

$$f(W^{cs}_{loc}) \cap B \subset W^{cs}_{loc}.$$

In addition, for any x *satisfying* $f^n(x) \in B$ *for all* $n \ge 0$,[2] *then* $x \in W^{cs}_{loc}$.

[2] $f^n(x)$ means the application of f on x repetitively for n times.

For each saddle point x^*, Theorem 7.3 implies the existence of a ball B_{x^*} centered at x^* and an invariant manifold $W^{cs}_{loc}(x^*)$ whose dimension is at most $n-1$. Let $B = \bigcup_{x^* \in \mathcal{X}} B_{x^*}$. With Lindelöf's Lemma (Lemma 7.6), there exists a countable $\mathcal{X}' \subset \mathcal{X}$ such that

$$B = \bigcup_{x^\star \in \mathcal{X}'} B_{x^*}.$$

Recall the dimension of $W^{cs}_{loc}(x^*)$ is at most $n-1$. Therefore $\mu(W^{cs}_{loc}(x^*)) = 0$. The measure of W^{cs}_{loc} can be subsequently bounded as

$$\mu(W^{cs}_{loc}) = \mu\left(\bigcup_{x^\star \in S'} W^{cs}_{loc}(x^\star)\right) \leq \sum_{x^* \in \mathcal{X}'} \mu\left(W^{cs}_{loc}(x^*)\right) = 0,$$

where the first inequality is from the semi-countable additivity of Lebesgue measure.

To relate the stable manifolds of these saddle points to the initialization, we need to analyze the gradient map. In contrast to previous analyses, we only show the gradient map is a *local* diffeomorphism instead of a global one, which is considerably weaker but sufficient for our purpose. This result is in the following lemma, which is proved in the appendix.

Lemma 7.2 *If a smooth map $g : \mathbb{R}^n \to \mathbb{R}^n$ is a local diffeomorphism, then for every open set S with $\mu(S) = 0$, the inverse set $g^{-1}(S)$ is also a zero-measure set, that is, $\mu\left(g^{-1}(S)\right) = 0$.*

Next, we show that we can choose a step size in $(0, 2/L)$ to make g a local diffeomorphism except for a zero-measure set.

Lemma 7.3 *The gradient map $g : \mathbb{R}^n \to \mathbb{R}^n$ in (7.6) is a local diffeomorphism in $\mathbb{R}^n$ for step sizes $h \in (0, 2/L)\backslash H$, where $H \subseteq [1/L, 2/L)$ has measure zero.*

Given Lemma 7.2 and Lemma 7.3, the rest of the proof is fairly straightforward. With Lemma 7.3, we know that under the step size $h \in (0, 2/L) \setminus H$ and $\mu(H) = 0$, gradient descent is a local diffeomorphism. Furthermore, with Lemma 7.2, we have

$$\mu\left(\bigcup_{i=0}^{\infty} g^{-i}(W^{cs}_{loc})\right) \leq \sum_{i=0}^{\infty} \mu(g^{-i}(W^{cs}_{loc})) = 0.$$

Thus, as long as the random initialization scheme is absolutely continuous with respect to the Lebesgue measure, GD will not converge to a saddle point.

7.6 Proof of Theorem 7.2

In this section we prove Theorem 7.2. First observe that if we can prove that a local manifold that converges to the strict saddle point has Lebesgue measure 0, then we can reuse the arguments for proving Theorem 7.1. To characterize the local behavior of GD with varying step sizes, we resort to a generalization of the seminal Hartman product map theorem.

7.6.1 Hartman Product Map Theorem

Before describing the theorem, we need to introduce some conditions and definitions.

Assumption 7.1 (Hypothesis (H_1) [Har71]) Let $(X, \|\cdot\|_X)$ and $(Y, \|\cdot\|_Y)$ be Banach spaces and $Z = X \times Y$ with norm $\|\cdot\|_Z = \max(\|\cdot\|_X, \|\cdot\|_Y)$. Define $Z_r(0) = \{z \in Z : \|z\|_Z < r\}$. Let $T_n(z) = (A_n x, B_n y) + (F_n(z), G_n(z))$ be a map from $Z_r(0)$ to Z with fixed point 0 and having a continuous Fréchet derivative. Let $A_n : X \to X$ and $B_n : Y \to Y$ be two linear operators and assume B_n is invertible. Suppose

$$\|A_n\|_X \leq a < 1, \quad \|B_n^{-1}\|_Y \leq 1/b \leq 1. \quad 0 < 4\delta < b - a, \quad 0 < a + 2\delta < 1, \tag{7.8}$$

$F_n(0) = 0$, $G_n(0) = 0$ and

$$\begin{cases} \|F_n(z_1) - F_n(z_2)\|_X \leq \delta \|z_1 - z_2\|_Z \\ \|G_n(z_1) - G_n(z_2)\|_Y \leq \delta \|z_1 - z_2\|_Z. \end{cases}$$

Here A_n represents local linear operator that acts on the space that corresponds to positive eigenvalues of the Hessian of a saddle point and B_n is a local linear operator that acts on the remaining space. F_n and G_n are higher order functions which vanish at 0.

The main mathematical object we study in this section is the following invariant set.

Definition 7.11 (Invariant Set) With the same notations in Assumption 7.1, let $T_1, \ldots, T_n$ be n maps from $Z_r(0)$ to Z and $S_n = T_n \circ T_{n-1} \circ \cdots \circ T_1$ be the product of the maps. Let $\mathcal{D}_n$ be the invariant set of the product operator S_n and $\mathcal{D} = \bigcap_{n=1}^{\infty} \mathcal{D}_n$.

This set corresponds to the points that will converge to the strict saddle point. To study its property, we consider a particular subset.

Definition 7.12 ([Har71])

$$\mathcal{D}^{a\delta} = \Big\{ z_0 = (x_0, y_0) \in \mathcal{D} : \; z_n \equiv S_n(z_0) \equiv (x_n, y_n) \text{ s.t. } \forall n, \|y_n\|_Y \leq \|x_n\|_X \leq (a + 2\delta)^n \|x_0\|_X \Big\}.$$

Now we are ready to state Hartman product map Theorem.

Theorem 7.4 (Theorem 7.1, [Har71]) *Under Assumption 7.1, the set $\mathcal{D}^{a\delta}$ is a C^1-manifold and satisfies $D^{a\delta} = \{z = (x, y) \in D | y = y_0(x)\}$ for some function y_0 which is continuous and has continuous Frèchet derivative $D_x y_0$ on $X_r(0)$ and $z_n = S_n(x_0, y_0) \equiv (x_n(x_0), y_n(x_0))$. Further, we have*

$$\|y_n(x^0) - y_n(x_0)\|_Y \leq \|x_n(x^0) - x_n(x_0)\|_X \leq (a + 2\delta)^n \|x^0 - x_0\|_X,$$
$$y_n(x_0) = y_0(x_n(x_0)),$$

for any $|x_0|, |x^0| < r$ and $n = 0, 1, \ldots$.

Remark 7.1 The C^1-manifold $y = y_0(x)$ is equivalent to $y - y_0(x) = 0$. The tangent manifold of y at the fixed point 0 is the intersection set $\bigcap_{i=1}^{dim(y)} \{(x, y) | \nabla_x y_i(x_0) \cdot x - y_i = 0\}$. In the $\mathbb{R}^n$ case, $\{(x, y) | \nabla_x y_i(x_0) \cdot x - y_i = 0\}$ is a subspace of $\mathbb{R}^n$ with dimension at most $n - 1$. Hence, its Lebesgue measure is 0.

Remark 7.2 Taking $x^0 = 0$ where 0 is a fixed point, we can rewrite the result of Theorem 7.4 as

$$\|y_n(x_0) - y_n(0)\|_Y \leq \|x_n(x_0) - x_n(0)\|_X \leq (a + 2\delta)^n \|x_0 - 0\|_X,$$
$$y_n(0) = y_0(x_n(0)) = 0, \quad x_n(0) = 0.$$

The following theorem from [Har71] implies that $D^{a\delta}$ is actually D.

Theorem 7.5 (Proposition 7.1, [Har71]) *Let $z_0 \in \mathcal{D}$; and $z_n = S_n(z_0)$ for $n = 0, 1, \ldots$.*

1. *If the inequality*

$$\|y_m\|_Y \geq \|x_m\|_X$$

holds for some $m \in \mathbb{N}$, then for $n > m$, we have

$$\|y_m\|_Y \geq \|x_m\|_X \qquad \|y_n\|_Y \geq (b - 2\delta)^{n-m} \|y_m\|_Y.$$

2. *Otherwise, for every $n \in \mathbb{N}$, we have*

$$\|y_n\|_Y \leq \|x_n\|_X \leq (a + 2\delta)^n \|x_0\|_X.$$

Using Theorem 7.5 we have the following useful corollary.

Corollary 7.2 *If $b - 2\delta > 1$, we have $D = D^{a\delta}$.*

7.6.2 Complete Proof of Theorem 7.2

We first correspond the parameters of GD to the notations in Assumption 7.1. Let x^* be a strict saddle point. Since $\nabla^2 f(x^*)$ is non-singular, it only contains positive and negative eigenvalues. We let $\mathbb{R}^n = X \times Y$, where X corresponds to the space of positive eigenvalues of $\nabla^2 f(x^*)$ and Y corresponds to the space of negative eigenvalues of $\nabla^2 f(x^*)$. For any $z \in \mathbb{R}^n$, we write $z = (x, y)$, where x represents the component in X and y represents the component in Y. Mappings $T_1, T_2, \ldots$ in Assumption 7.1 correspond to the gradient maps. A_n,B_n,F_n, and G_n are thus defined accordingly. The next lemma shows under our assumption on the step size, GD dynamics satisfies Assumption 7.1.

Lemma 7.4 *Suppose $f \in C^2(\mathbb{R}^n)$ with Hessian Lipschitz constant K and x^* a strict saddle point with $L = \|\nabla^2 f(x^*)\|_2$ and $\mu = \|\left(\nabla^2 f(x^*)\right)^{-1}\|_2^{-1}$. For any fixed $\epsilon_0 \in (0, 1)$, if the step size satisfies $h_t \in \left[\frac{\epsilon_0}{L}, \frac{2-\epsilon_0}{L}\right]$, we have*

$$\|A_t\|_2 \leq 1 - \epsilon_0 \qquad \text{and} \qquad \|B_t^{-1}\|_2 \leq \frac{1}{1 + \frac{\epsilon_0 \mu}{L}}$$

and for any $z_1, z_2 \in \mathbb{R}^n$.

$$\max(\|F_t(z_1) - F_t(z_2)\|_2, \|G_t(z_1) - G_t(z_2)\|_2) \leq \delta \|z_1 - z_2\|_2,$$

where $\delta = \frac{\epsilon}{5}$ and $r = \frac{\epsilon L}{20K}$ (c.f. Assumption 7.1).

Let D be the invariant set defined in Definition 7.12. From Theorem 7.4, Remarks 7.1, and 7.2, we know that the induced shrinking C^1 manifold $\mathcal{D}^{a\delta}$ defined in Definition 7.12 has dimension at most $n - 1$. Furthermore, by Corollary 7.2 we know that $\mathcal{D} = \mathcal{D}^{a\delta}$. Therefore, the set of points converging to the strict saddle point has zero Lebesgue measure. Similar to the proof of Theorem 7.1, since the gradient map is a local diffeomorphism, we can see that with random initialization, GD will not converge to any saddle point. The proof is complete.

7.7 Additional Theorems

Lemma 7.5 (The Inverse Function Theorem) *Let $f : M \to N$ be a smooth map, and $\dim(M) = \dim(N)$. Suppose that the Jacobian Df_p is non-singular at some $p \in M$. Then f is a local diffeomorphism at p, i.e., there exists an open neighborhood U of p such that*

1. *f is one-to-one on U.*
2. *$f(U)$ is open in N.*
3. *$f^{-1} : f(U) \to U$ is smooth.*

In particular, $D(f^{-1})_{f(p)} = (Df_p)^{-1}$.

Lemma 7.6 (Lindelöf's Lemma) *For every open cover there is a countable subcover.*

7.8 Technical Proofs

Proof Proof of Lemma 7.2 With Theorem 7.5, we know that for every $x \in S$, there exists an open set $U_x \in \mathbb{R}^n$ such that g is non-singular. Let $W_x = S \cap U_x$, then we have

$$S \subseteq \bigcup_{x \in S} W_x.$$

With Lindelöf's lemma, there exists a set S' with countable elements x such that

$$S \subseteq \bigcup_{x \in S'} W_x.$$

Since the Dg is non-singular on W_x, we know that g on W_x is one-one-onto. Hence, we have $\mu(g^{-1}(W_x)) = 0$. Hence, we have

$$\mu\left(g^{-1}\left(\bigcup_{x \in S} W_x\right)\right) = \mu\left(g^{-1}\left(\bigcup_{x \in S'} W_x\right)\right) \leq \mu\left(\bigcup_{x \in S'} g^{-1}(x)\right)$$
$$\leq \sum_{x \in S'} \mu(g^{-1}(x)) = 0,$$

where the second inequality is from the monotony of Lebesgue measure and the third inequality is from the semi-countable additivity of Lebesgue measure. □

Proof Proof of Lemma 7.3 If the Jacobian of the gradient map Dg is non-singular at some point $x \in \mathbb{R}^n$, with the continuity of the Jacobian Dg, we know that Dg is non-singular at some open neighborhood $\mathcal{U}_x$ of the point x. Hence, we have

$$\mathbb{R}^n \subseteq \bigcup_{x \in \mathbb{R}^n} \mathcal{U}_x.$$

With Lindelöf's lemma, there exists a set $\mathcal{S}$ with countable number of $x \in \mathbb{R}^n$ such that

$$\mathbb{R}^n \subseteq \bigcup_{x \in \mathcal{S}} \mathcal{U}_x$$

Let $\mathcal{H}_x$ be the step size that Jacobian Dg is singular at the open set $\mathcal{U}_x$. With the definition of $\mathcal{U}_x$, we know that there are at most n elements in $\mathcal{H}_x$. Hence, we have

$$\mu(H_x) = 0 \qquad \text{and} \qquad H = \bigcup_{x \in \mathcal{S}} H_x,$$

where H satisfies that the Jacobian of the gradient map is non-singular with step size $h \in \left(0, \frac{2}{L}\right) \setminus H$. With the semi-countable additivity of Lebesgue measure, we have

$$\mu(H) = \mu\left(\bigcup_{x \in \mathcal{S}} H_x\right) \le \sum_{x \in \mathcal{S}} \mu(H_x) = 0.$$

□

Proof Proof of Proposition 7.1 Consider the following quadratic function:

$$f(x) = \frac{1}{2} x^T A x,$$

where A is a diagonal matrix $A = \mathbf{diag}(\lambda_1, \ldots, \lambda_n)$ and satisfies $\lambda_1 > \lambda_2 > \ldots, \lambda_n > 0$. The global gradient Lipschitz constant L of f is λ_1. Now consider the gradient dynamics

$$x_{t+1} = x_t - hAx_t = (I - hA)\, x_t.$$

Since $h \ge \frac{2}{L}$, $\lambda_{\max}(I - hA) \ge 1$. Therefore, the sequence $\{x_0, x_1, \ldots\}$ does not converge. □

Proof Proof of Lemma 7.4 If $x_t \in V_r(x^\star)$, the step size satisfies

$$h_t \in \left[\frac{\epsilon_0}{L_{(x_t,r)}}, \frac{2-\epsilon_0}{L_{(x_t,r)}}\right] \subseteq \left[\frac{\epsilon_0}{L - 2Kr}, \frac{2-\epsilon_0}{L}\right] \subseteq \left[\frac{\epsilon_0}{L(1-0.1\epsilon_0)}, \frac{2-\epsilon_0}{L}\right].$$

Therefore, we know

$$h_t \in \left[\frac{\epsilon_0'}{L}, \frac{2-\epsilon_0'}{L}\right],$$

where $\epsilon_0' = \frac{\epsilon_0}{1-0.1\epsilon_0'}$. Since both A_t and B_t are diagonal, then the 2-norm is equal to the maximum eigenvalue, that is,

$$\|A_t\|_2 = 1 - \max|\lambda(\nabla^2 | f(x^\star)) \cdot h \le 1 - \epsilon_0$$

$$\|B_t^{-1}\|_2 = \frac{1}{1 + \mu \min|\lambda(\nabla^2 f(x^*))|h} \le \frac{1}{1 + \frac{\epsilon_0 \mu}{L}}.$$

Furthermore, we have

$$
\begin{aligned}
&\max(\|F_1(z_1) - F_1(z_2)\|_2, \|F_2(z_1) - F_2(z_2)\|_2) \\
&\leq h\|(\nabla f(x) - \nabla f(y)) + \nabla^2 f(x^\star)(x - y)\|_2 \\
&\leq hK\left(\|z_1\|_2 + \|z_2\|_2\right)\|z_1 - z_2\|_2.
\end{aligned}
$$

Plugging in our assumption on the step size we have the desired result. □

7.9 Conclusions

In this chapter we considered optimal and adaptive step-size rules for gradient descent (GD) applied to *non-convex* optimization problems. We proved that GD with fixed step sizes not exceeding $2/L$ will not converge to strict saddle points almost surely, generalizing previous works of [LSJR16, PP16] that require step sizes to not exceed $1/L$. We also establish escaping strict saddle point properties of GD under varying/adaptive step sizes under additional conditions.

One important open question is to derive explicit *rates of convergence* for the GD algorithm with different step-size rules for non-convex objective functions. It is particularly interesting to study non-convex problems for which GD converges to local minima with number of iterations *polynomial* in problem dimension d. While the work of [DJL$^+$17] rules out such possibility for general smooth f, polynomial iteration complexity of GD might still be possible for non-convex objectives under additional assumptions.

Chapter 8
A Conservation Law Method Based on Optimization

This chapter is organized as follows: In Sect. 8.1, we warm up with an analytical solution for simple 1-D quadratic function. In Sect. 8.2, we propose the artificially dissipating energy algorithm, energy conservation algorithm, and the combined algorithm based on the symplectic Euler scheme, and remark a second-order scheme—the Störmer–Verlet scheme. In Sect. 8.3, we propose the locally theoretical analysis for high-speed convergence. Section 8.4 proposes the experimental demonstration. In Sect. 8.4, we propose the experimental result for the proposed algorithms on strongly convex, non-strongly convex, and non-convex functions in high dimension. Finally, we propose some perspective view for the proposed algorithms and two adventurous ideas based on the evolution of Newton's second law—fluid and quantum.

8.1 Warm-up: An Analytical Demonstration for Intuition

For a simple 1-D function with ill-conditioned Hessian, $f(x) = \frac{1}{200}x^2$ with the initial position at $x_0 = 1000$. The solution and the function value along the solution for (3.9) are given by

$$\begin{cases} x(t) = x_0 e^{-\frac{1}{100}t} & (8.1) \\ f(x(t)) = \dfrac{1}{200} x_0^2 e^{-\frac{1}{50}t}. & (8.2) \end{cases}$$

Parts of this chapter is in the paper titled "A Conservation Law Method in Optimization" by Bin Shi et al. (2017) published by 10th NIPS Workshop on Optimization for Machine Learning.

B. Shi, S. S. Iyengar, *Mathematical Theories of Machine Learning - Theory and Applications*, https://doi.org/10.1007/978-3-030-17076-9_8

The solution and the function value along the solution for (3.10) with the optimal friction parameter $\gamma_t = \frac{1}{5}$ are

$$\begin{cases} x(t) = x_0\left(1 + \dfrac{1}{10}t\right)e^{-\frac{1}{10}t} & (8.3)\\ f(x(t)) = \dfrac{1}{200}x_0^2\left(1+\dfrac{1}{10}t\right)^2 e^{-\frac{1}{5}t}. & (8.4)\end{cases}$$

The solution and the function value along the solution for (3.12) are

$$\begin{cases} x(t) = x_0\cos\left(\dfrac{1}{10}t\right) \quad \text{and} \quad v(t) = x_0 \sin\left(\dfrac{1}{10}t\right) & (8.5)\\ f(x(t)) = \dfrac{1}{200}x_0^2\cos^2\left(\dfrac{1}{10}t\right), & (8.6)\end{cases}$$

stop at the point that $|v|$ arrives maximum. Combined with (8.2), (8.4), and (8.6) with stop at the point that $|v|$ arrives maximum, the function value approximating $f(x^\star)$ is shown as below.

From the analytical solution for local convex quadratic function with maximum eigenvalue L and minimum eigenvalue μ, in general, the step size by $\frac{1}{\sqrt{L}}$ for momentum method and Nesterov's accelerated gradient method, hence the simple estimate for iterative times is approximately

$$n \sim \frac{\pi}{2}\sqrt{\frac{L}{\mu}}.$$

Hence, the iterative times n is proportional to the reciprocal of the square root of minimal eigenvalue $\sqrt{\mu}$, which is essentially different from the convergence rate of the gradient method and momentum method (Fig. 8.1).

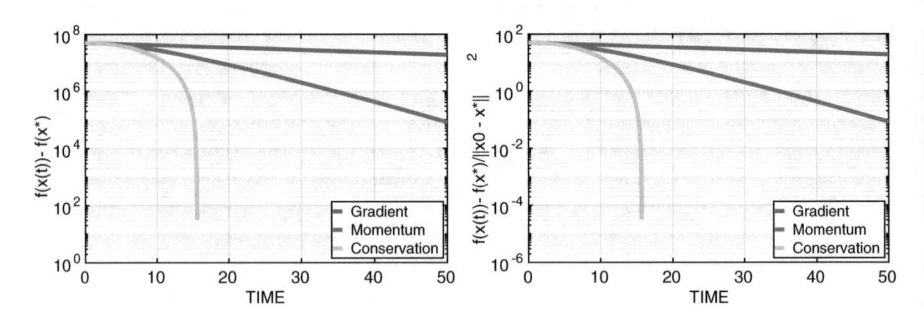

Fig. 8.1 Minimizing $f(x) = \frac{1}{200}x^2$ by the analytical solution for (8.2), (8.4), and (8.6) with stop at the point that $|v|$ arrives maximum, starting from $x_0 = 1000$ and the numerical step size $\Delta t = 0.01$

8.2 Symplectic Scheme and Algorithms

In this chapter, we utilize the first-order symplectic Euler scheme from numerically solving Hamiltonian system as below to propose the corresponding artificially dissipating energy algorithm to find the global minima for convex function, or local minima in non-convex function

$$\begin{cases} x_{k+1} = x_k + hv_{k+1} \\ v_{k+1} = v_k - h\nabla f(x_k). \end{cases} \tag{8.7}$$

Then by the observability of the velocity, we propose the energy conservation algorithm for detecting local minima along the trajectory. Finally, we propose a combined algorithm to find better local minima between some local minima.

Remark 8.1 In all the algorithms below, the symplectic Euler scheme can be replaced by the Störmer–Verlet scheme

$$\begin{cases} v_{k+1/2} = v_k - \dfrac{h}{2}\nabla f(x_k) \\ x_{k+1} = x_k + hv_{k+1/2} \\ v_{k+1} = v_{k+1/2} - \dfrac{h}{2}\nabla f(x_{k+1}). \end{cases} \tag{8.8}$$

This works better than the symplectic scheme even if doubling step size and keeping the left–right symmetry of the Hamiltonian system. The Störmer–Verlet scheme is the natural discretization for 2nd-order ODE which is named as leap-frog scheme in PDEs

$$x_{k+1} - 2x_k + x_{k-1} = -h^2\nabla f(x_k). \tag{8.9}$$

We remark that the discrete scheme (8.9) is different from the finite difference approximation by the forward Euler method to analyze the stability of 2nd ODE in [SBC14], since the momentum term is biased.

8.2.1 Artificially Dissipating Energy Algorithm

Firstly, the artificially dissipating energy algorithm based on (8.7) is proposed as below.

Algorithm 1 Artificially dissipating energy algorithm

1: Given a starting point $x_0 \in \mathbf{dom}(f)$
2: Initialize the step length h, maxiter, and the velocity variable $v_0 = 0$
3: Initialize the iterative variable $v_{iter} = v_0$
4: **while** $\|\nabla f(x)\| > \epsilon$ and $k <$ maxiter **do**
5: Compute v_{iter} from the below equation in (8.7)
6: **if** $\|v_{iter}\| \leq \|v\|$ **then**
7: $v = 0$
8: **else**
9: $v = v_{iter}$
10: **end if**
11: Compute x from the above equation in (8.7)
12: $x_k = x$;
13: $f(x_k) = f(x)$;
14: $k = k + 1$;
15: **end while**

Remark 8.2 In the actual Algorithm 1, the codes in line 15 and 16 are not needed in the while loop in order to speed up the computation.

A Simple Example for Illustration

Here, we use a simple convex quadratic function with ill-conditioned eigenvalue for illustration as below:

$$f(x_1, x_2) = \frac{1}{2}\left(x_1^2 + \alpha x_2^2\right), \tag{8.10}$$

of which the maximum eigenvalue is $L = 1$ and the minimum eigenvalue is $\mu = \alpha$. Hence the scale of the step size for (8.10) is

$$\frac{1}{L} = \sqrt{\frac{1}{L}} = 1.$$

In Fig. 8.2, we demonstrate the convergence rate of gradient method, momentum method, Nesterov's accelerated gradient method, and artificially dissipating energy method with the common step size $h = 0.1$ and $h = 0.5$, where the optimal friction parameter for momentum method $\gamma = \frac{1-\sqrt{\alpha}}{1+\sqrt{\alpha}}$ with $\alpha = 10^{-5}$. A further result for comparison with the optimal step size in gradient method $h = \frac{2}{1+\alpha}$, the momentum method $h = \frac{4}{(1+\sqrt{\alpha})^2}$, and Nesterov's accelerated gradient method with $h = 1$ and the artificially dissipating energy method with $h = 0.5$ is shown in Fig. 8.3.

With the illustrative convergence rate, we need to learn the trajectory. Since the trajectories of all the four methods are so narrow in ill-condition function in (8.10), we use a relatively good-conditioned function to show it as $\alpha = \frac{1}{10}$ in Fig. 8.4.

The fact highlighted in Fig. 8.4 demonstrates the gradient correction decreases the oscillation when compared with the momentum method. A clearer observation

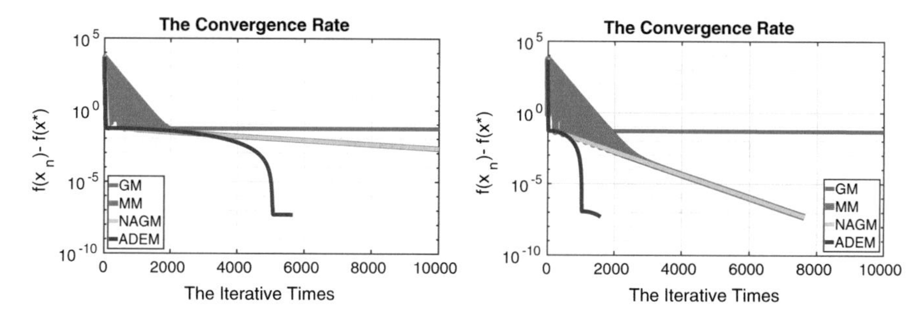

Fig. 8.2 Minimizes the function in (8.10) for artificially dissipating energy algorithm comparing with gradient method, momentum method, and Nesterov's accelerated gradient method with stop criteria $\epsilon = 1e-6$. The step size: Left: $h = 0.1$; Right: $h = 0.5$

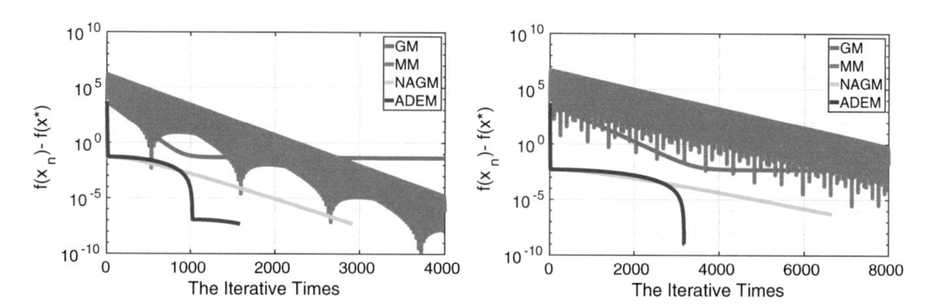

Fig. 8.3 Minimizes the function in (8.10) for artificially dissipating energy algorithm comparing with gradient method, momentum method, and Nesterov's accelerated gradient method with stop criteria $\epsilon = 1e-6$. The Coefficient α: Left: $\alpha = 10^{-5}$; Right: $\alpha = 10^{-6}$

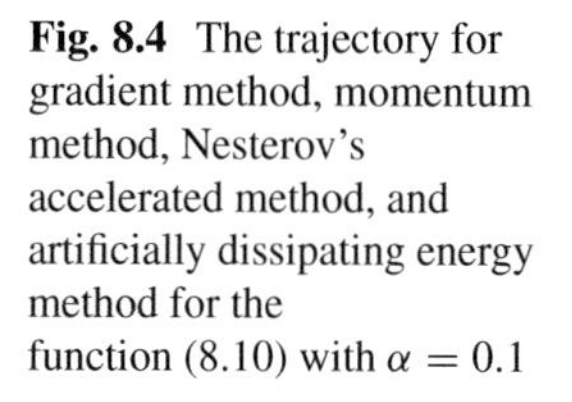
Fig. 8.4 The trajectory for gradient method, momentum method, Nesterov's accelerated method, and artificially dissipating energy method for the function (8.10) with $\alpha = 0.1$

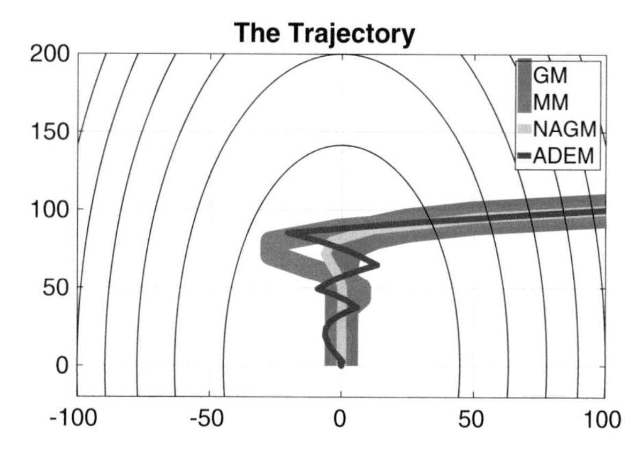

is the artificially dissipating method shares the same property with the other three methods by the law of nature, that is, if the trajectory comes into the local minima in one dimension it will not leave it very far. However, from Figs. 8.2 and 8.3, we see the more rapid convergence rate from using the artificially dissipating energy method.

8.2.2 *Detecting Local Minima Using Energy Conservation Algorithm*

Here, the energy conservation algorithm based on (8.7) is proposed as below.

Algorithm 2 Energy conservation algorithm

1: Given a starting point $x_0 \in \mathbf{dom}(f)$
2: Initialize the step size h and the maxiter
3: Initialize the velocity $v_0 > 0$ and compute $f(x_0)$
4: Compute the velocity x_1 and v_1 from Eq. (8.7), and compute $f(x_1)$
5: **for** $k = 1 : n$ **do**
6: Compute x_{k+1} and v_{k+1} from (8.7)
7: Compute $f(x_{k+1})$
8: **if** $\|v_k\| \geq \|v_{k+1}\|$ and $\|v_k\| \geq \|v_{k-1}\|$ **then**
9: Record the position x_k
10: **end if**
11: **end for**

Remark 8.3 In Algorithm 2, we can set $v_0 > 0$ so that the total energy is large enough to climb up some high peak. Similar to Algorithm 1 defined earlier, the function value $f(x)$ is not need in the while loop in order to speed up the computation.

The Simple Example for Illustration

Here, we use the non-convex function for illustration as below:

$$f(x) = \begin{cases} 2\cos(x), & x \in [0, 2\pi] \\ \cos(x) + 1, & x \in [2\pi, 4\pi] \\ 3\cos(x) - 1, & x \in [4\pi, 6\pi], \end{cases} \tag{8.11}$$

which is the 2nd-order smooth function but not 3rd-order smooth. The maximum eigenvalue can be calculated as below:

$$\max_{x \in [0, 6\pi]} |f''(x)| = 3.$$

The step length is set $h \sim \sqrt{\frac{1}{L}}$. We illustrate that Algorithm 2 simulates the trajectory and find the local minima in Fig. 8.5.

Another 2D potential function is shown as below:

$$f(x_1, x_2) = \frac{1}{2}\left[(x_1 - 4)^2 + (x_2 - 4)^2 + 8\sin(x_1 + 2x_2)\right], \tag{8.12}$$

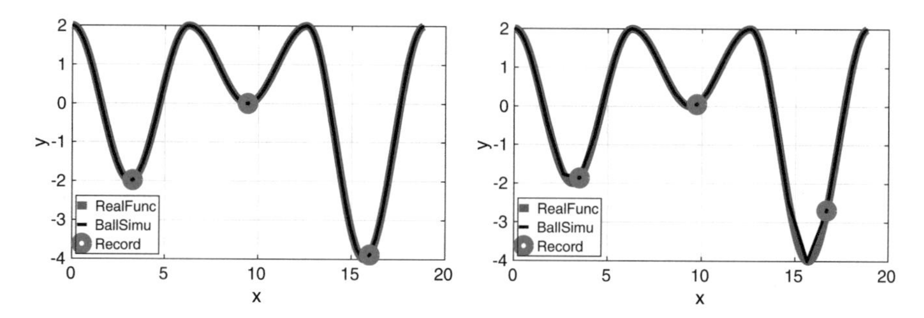

Fig. 8.5 Left: the step size $h = 0.1$ with 180 iterative times. Right: the step size $h = 0.3$ with 61 iterative times

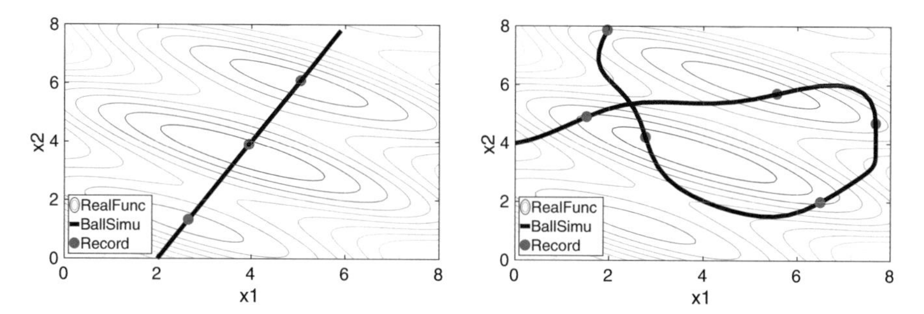

Fig. 8.6 The common step size is set $h = 0.1$. Left: the position at (2, 0) with 23 iterative times. Right: the position at (0, 4) with 62 iterative times

which is the smooth function with domain in $(x_1, x_2) \in [0, 8] \times [0, 8]$. The maximum eigenvalue can be calculated as below:

$$\max_{x \in [0, 6\pi]} |\lambda(f''(x))| \geq 16.$$

The step length is set $h \sim \sqrt{\frac{1}{L}}$. We illustrate that Algorithm 2 simulates the trajectory and find the local minima as in Fig. 8.6.

Remark 8.4 We point out that the energy conservation algorithm for detecting local minima along the trajectory cannot detect the saddle point in the sense of almost every, since the saddle point in the original function $f(x)$ is also a saddle point for the energy function $H(x, v) = \frac{1}{2}\|v\|^2 + f(x)$. The proof process is fully the same in [LSJR16].

8.2.3 Combined Algorithm

Finally, we propose the comprehensive algorithm combining the artificially dissipating energy algorithm (Algorithm 1) and the energy conservation algorithm (2) to find global minima.

Algorithm 3 Combined algorithm

1: Given some starting points $x_{0,i} \in \textbf{dom}(f)$ with $i = 1, \ldots, n$
2: Implement algorithm 2 detecting the position there exists local minima, noted as x_j with $j = 1, \ldots, m$
3: Implement algorithm 1 from the result on line 2 finding the local minima, noted as x_k with $k = 1, \ldots, l$
4: Comparison of $f(x_k)$ with $k = 1, \ldots, l$ to find global minima.

Remark 8.5 We remark that the combined algorithm (Algorithm 3) cannot guarantee to find global minima if the initial position is not ergodic. The tracking local minima is dependent on the trajectory. However, the time of computation and precision based on the proposed algorithm is far better than the large sampled gradient method. Our proposed algorithm first makes the identified global minima to become possible.

8.3 An Asymptotic Analysis for the Phenomena of Local High-Speed Convergence

In this section, we analyze the phenomena of high-speed convergence shown in Figs. 8.1, 8.2, and 8.3. Without loss of generality, we use the translate transformation $y_k = x_k - x^\star$ ($x^\star$ is the point of local minima) and $v_k = v_k$ into (8.7), shown as below:

$$\begin{cases} y_{k+1} = y_k + hv_{k+1} \\ v_{k+1} = v_k - h\nabla f(x^\star + y_k), \end{cases} \tag{8.13}$$

the locally linearized scheme of which is given as below:

$$\begin{cases} y_{k+1} = y_k + hv_{k+1} \\ v_{k+1} = v_k - h\nabla^2 f(x^\star)y_k. \end{cases} \tag{8.14}$$

Remark 8.6 The local linearized analysis is based on the stability theorem in finite dimension, the invariant stable manifold theorem, and Hartman–Grobman linearized map theorem [Har82]. The thought is firstly used in [Pol64] to estimate the local convergence of momentum method. And in the paper [LSJR16], the thought is used to exclude the possibility of convergence to saddle point. However, the two theorems above belong to the qualitative theorem of ODE. Hence, the linearized scheme (8.14) is only an approximate estimate for the original scheme (8.13) locally.

8.3.1 Some Lemmas for the Linearized Scheme

Let A be the positive-semidefinite and symmetric matrix to represent $\nabla^2 f(x^\star)$ in (8.14).

Lemma 8.1 *The numerical scheme shown as below*

$$\begin{pmatrix} x_{k+1} \\ v_{k+1} \end{pmatrix} = \begin{pmatrix} I - h^2 A & hI \\ -hA & I \end{pmatrix} \begin{pmatrix} x_k \\ v_k \end{pmatrix} \tag{8.15}$$

is equivalent to the linearized symplectic Euler scheme (8.14), where we note that the linear transformation is

$$M = \begin{pmatrix} I - h^2 A & hI \\ -hA & I \end{pmatrix}. \tag{8.16}$$

Proof

$$\begin{pmatrix} I & -hI \\ 0 & I \end{pmatrix} \begin{pmatrix} x_{k+1} \\ v_{k+1} \end{pmatrix} = \begin{pmatrix} I & 0 \\ -hA & I \end{pmatrix} \begin{pmatrix} x_k \\ v_k \end{pmatrix} \Leftrightarrow \begin{pmatrix} x_{k+1} \\ v_{k+1} \end{pmatrix} = \begin{pmatrix} I - h^2 A & hI \\ -hA & I \end{pmatrix} \begin{pmatrix} x_k \\ v_k \end{pmatrix}$$

□

Lemma 8.2 *For every $2n \times 2n$ matrix M in (8.16), there exists the orthogonal transformation $U_{2n\times 2n}$ such that the matrix M is similar as below:*

$$U^T M U = \begin{pmatrix} T_1 & & & \\ & T_2 & & \\ & & \ddots & \\ & & & T_n, \end{pmatrix} \tag{8.17}$$

where T_i $(i = 1, \ldots, n)$ is a 2×2 matrix with the form

$$T_i = \begin{pmatrix} 1 - \omega_i^2 h^2 & h \\ -\omega_i^2 h & 1, \end{pmatrix} \tag{8.18}$$

where ω_i^2 is the eigenvalue of the matrix A.

Proof Let Λ be the diagonal matrix with the eigenvalues of the matrix A as below:

$$\Lambda = \begin{pmatrix} \omega_1^2 & & & \\ & \omega_2^2 & & \\ & & \ddots & \\ & & & \omega_n^2 \end{pmatrix}.$$

Since A is positive define and symmetric, there exists orthogonal matrix U_1 such that

$$U_1^T A U_1 = \Lambda.$$

Let Π be the permutation matrix satisfying

$$\Pi_{i,j} = \begin{cases} 1, & j \text{ odd}, \ i = \dfrac{j+1}{2} \\ 1, & j \text{ even}, \ i = n + \dfrac{j}{2} \\ 0, & \text{otherwise}, \end{cases}$$

where i is the row index and j is the column index. Then, let $U = \mathbf{diag}(U_1, U_1)\Pi$, we have by conjugation

$$\begin{aligned} U^T M U &= \Pi^T \begin{pmatrix} U_1^T & \\ & U_1^T \end{pmatrix} \begin{pmatrix} I - h^2 A & hI \\ -hA & I \end{pmatrix} \begin{pmatrix} U_1 & \\ & U_1 \end{pmatrix} \Pi \\ &= \Pi^T \begin{pmatrix} I - h^2 \Lambda & hI \\ -h\Lambda & I \end{pmatrix} \Pi \\ &= \begin{pmatrix} T_1 & & & \\ & T_2 & & \\ & & \ddots & \\ & & & T_n \end{pmatrix}. \end{aligned}$$

□

From Lemma 8.2, we know that Eq. (8.15) can be written as the equivalent form

$$\begin{pmatrix} (U_1^T x)_{k+1,i} \\ (U_1^T v)_{k+1,i} \end{pmatrix} = T_i \begin{pmatrix} (U_1^T x)_{k,i} \\ (U_1^T v)_{k,i} \end{pmatrix} = \begin{pmatrix} 1 - \omega_i^2 h^2 & h \\ -\omega_i^2 h & 1 \end{pmatrix} \begin{pmatrix} (U_1^T x)_{k,i} \\ (U_1^T v)_{k,i} \end{pmatrix}, \tag{8.19}$$

where $i = 1, \ldots, n$.

Lemma 8.3 *For any step size h satisfying $0 < h\omega_i < 2$, the eigenvalues of the matrix T_i are complex with absolute value* 1.

Proof For $i = 1, \dots, n$, we have

$$|\lambda I - T_i| = 0 \Leftrightarrow \lambda_{1,2} = 1 - \frac{h^2\omega_i^2}{2} \pm h\omega_i \sqrt{1 - \frac{h^2\omega_i^2}{4}}.$$

□

Let θ_i and ϕ_i for $i = 1, \dots, n$ for the new coordinate variables be as follows:

$$\begin{cases} \cos\theta_i = 1 - \dfrac{h^2\omega_i^2}{2} \\ \sin\theta_i = h\omega_i \sqrt{1 - \dfrac{h^2\omega_i^2}{4}} \end{cases}, \qquad \begin{cases} \cos\phi_i = \dfrac{h\omega_i}{2} \\ \sin\phi_i = \sqrt{1 - \dfrac{h^2\omega_i^2}{4}}. \end{cases} \tag{8.20}$$

In order to make θ_i and ϕ_i located in $\left(0, \frac{\pi}{2}\right)$, we need to shrink to $0 < h\omega_i < \sqrt{2}$.

Lemma 8.4 *With the new coordinate in (8.20) for $0 < h\omega_i < \sqrt{2}$, we have*

$$2\phi_i + \theta_i = \pi \tag{8.21}$$

and

$$\begin{cases} \sin\theta_i = \sin(2\phi_i) = h\omega_i \sin\phi_i \\ \sin(3\phi_i) = -\left(1 - h^2\omega_i^2\right)\sin\phi_i. \end{cases} \tag{8.22}$$

Proof With sum–product identities of trigonometric function, we have

$$\begin{aligned} \sin(\theta_i + \phi_i) &= \sin\theta_i \cos\phi_i + \cos\theta_i \sin\phi_i \\ &= h\omega_i \sqrt{1 - \frac{h^2\omega_i^2}{4}} \cdot \frac{h\omega_i}{2} + \left(1 - \frac{h^2\omega_i^2}{2}\right)\sqrt{1 - \frac{h^2\omega_i^2}{4}} \\ &= \sqrt{1 - \frac{h^2\omega_i^2}{4}} \\ &= \sin\phi_i. \end{aligned}$$

Since $0 < h\omega_i < 2$, we have $\theta_i, \phi_i \in \left(0, \frac{\pi}{2}\right)$, we can obtain that

$$\theta_i + \phi_i = \pi - \phi_i \Leftrightarrow \theta_i = \pi - 2\phi_i$$

and with the coordinate transformation in (8.20), we have

$$\sin\theta_i = h\omega_i \sin\phi_i \Leftrightarrow \sin(2\phi_i) = h\omega_i \sin\phi_i.$$

Next, we use sum–product identities of trigonometric function furthermore

$$\begin{aligned}\sin(\theta_i - \phi_i) &= \sin\theta_i \cos\phi_i - \cos\theta_i \sin\phi_i \\ &= h\omega_i\sqrt{1 - \frac{h^2\omega_i^2}{4}} \cdot \frac{h\omega_i}{2} - \left(1 - \frac{h^2\omega_i^2}{2}\right)\sqrt{1 - \frac{h^2\omega_i^2}{4}} \\ &= \left(h^2\omega_i^2 - 1\right)\sqrt{1 - \frac{h^2\omega_i^2}{4}} \\ &= -\left(1 - h^2\omega_i^2\right)\sin\phi_i\end{aligned}$$

and with $\theta_i = \pi - 2\phi_i$, we have

$$\sin(3\phi_i) = -\left(1 - h^2\omega_i^2\right)\sin\phi_i.$$

□

Lemma 8.5 *With the new coordinate in (8.20), the matrix T_i $(i = 1, \ldots, n)$ in (8.18) can expressed as below:*

$$T_i = \frac{1}{\omega_i\left(e^{-i\phi_i} - e^{i\phi_i}\right)}\begin{pmatrix} 1 & 1 \\ \omega_i e^{i\phi_i} & \omega_i e^{-i\phi_i}\end{pmatrix}\begin{pmatrix} e^{i\theta_i} & 0 \\ 0 & e^{-i\theta_i}\end{pmatrix}\begin{pmatrix} \omega_i e^{-i\phi_i} & -1 \\ -\omega_i e^{i\phi_i} & 1\end{pmatrix}. \tag{8.23}$$

Proof For the coordinate transformation in (8.20), we have

$$T_i\begin{pmatrix} 1 \\ \omega_i e^{i\phi_i}\end{pmatrix} = \begin{pmatrix} 1 \\ \omega_i e^{i\phi_i}\end{pmatrix} e^{i\theta_i} \quad \text{and} \quad T_i\begin{pmatrix} 1 \\ \omega_i e^{-i\phi_i}\end{pmatrix} = \begin{pmatrix} 1 \\ \omega_i e^{-i\phi_i}\end{pmatrix} e^{-i\theta_i}.$$

Hence, (8.23) is proved. □

8.3.2 *The Asymptotic Analysis*

Theorem 8.1 *Let the initial value x_0 and v_0, after the first k steps without resetting the velocity, the iterative solution (8.14) with the equivalent form (8.19) has the form as below:*

$$\begin{pmatrix} (U_1^T x)_{k,i} \\ (U_1^T v)_{k,i} \end{pmatrix} = T_i^k \begin{pmatrix} (U_1^T x)_{0,i} \\ (U_1^T v)_{0,i} \end{pmatrix} = \begin{pmatrix} -\frac{\sin(k\theta_i - \phi_i)}{\sin\phi_i} & \frac{\sin(k\theta_i)}{\omega_i \sin\phi_i} \\ -\frac{\omega_i \sin(k\theta_i)}{\sin\phi_i} & \frac{\sin(k\theta_i+\phi_i)}{\sin\phi_i} \end{pmatrix} \begin{pmatrix} (U_1^T x)_{0,i} \\ (U_1^T v)_{0,i} \end{pmatrix}. \tag{8.24}$$

Proof With Lemma 8.5 and the coordinate transformation (8.20), we have

$$\begin{aligned} T_i^k &= \frac{1}{\omega_i\left(e^{-i\phi_i} - e^{i\phi_i}\right)} \begin{pmatrix} 1 & 1 \\ \omega_i e^{i\phi_i} & \omega_i e^{-i\phi_i} \end{pmatrix} \begin{pmatrix} e^{i\theta_i} & 0 \\ 0 & e^{-i\theta_i} \end{pmatrix}^k \begin{pmatrix} \omega_i e^{-i\phi_i} & -1 \\ -\omega_i e^{i\phi_i} & 1 \end{pmatrix} \\ &= \frac{1}{\omega_i\left(e^{-i\phi_i} - e^{i\phi_i}\right)} \begin{pmatrix} 1 & 1 \\ \omega_i e^{i\phi_i} & \omega_i e^{-i\phi_i} \end{pmatrix} \begin{pmatrix} \omega e^{i(k\theta_i - \phi_i)} & -e^{ik\theta_i} \\ -\omega e^{-i(k\theta_i-\phi_i)} & e^{-ik\theta_i} \end{pmatrix} \\ &= \begin{pmatrix} -\frac{\sin(k\theta_i - \phi_i)}{\sin\phi_i} & \frac{\sin(k\theta_i)}{\omega_i \sin\phi_i} \\ -\frac{\omega_i \sin(k\theta_i)}{\sin\phi_i} & \frac{\sin(k\theta_i+\phi_i)}{\sin\phi_i} \end{pmatrix}. \end{aligned}$$

The proof is complete. □

Comparing (8.24) and (8.19), we can obtain that

$$\frac{\sin(k\theta_i - \phi_i)}{\sin\phi_i} = 1 - h^2\omega_i^2.$$

With the initial value $(x_0, 0)^T$, then the initial value for (8.19) is $(U_1^T x_0, 0)$. In order to make sure the numerical solution or the iterative solution owns the same behavior as the analytical solution, we need to set $0 < h\omega_i < 1$.

Remark 8.7 Here, the behavior is similar as the thought in [LSJR16]. The step size $0 < hL < 2$ makes sure the global convergence of gradient method. And the step size $0 < hL < 1$ makes the uniqueness of the trajectory along the gradient method, the thought of which is equivalent of the existence and uniqueness of the solution for ODE. Actually, the step size $0 < hL < 1$ owns the property with the solution of ODE, the continuous-limit version. A global existence of the solution for gradient system is proved in [Per13].

For the good-conditioned eigenvalue of the Hessian $\nabla^2 f(x^\star)$, every method such as gradient method, momentum method, Nesterov's accelerated gradient method, and artificially dissipating energy method has the good convergence rate shown by the experiment. However, for our artificially dissipating energy method, since there are trigonometric functions from (8.24), we cannot propose the rigorous mathematic proof for the convergence rate. If everybody can propose a theoretical proof, it is very beautiful. Here, we propose a theoretical approximation for ill-conditioned case, that is, the direction with small eigenvalue $\lambda(\nabla^2 f(x^\star)) \ll L$.

Assumption 8.1 If the step size $h = \frac{1}{\sqrt{L}}$ for (8.14), for the ill-conditioned eigenvalue $\omega_i \ll \sqrt{L}$, the coordinate variable can be approximated by the analytical solution as

$$\theta_i = h\omega_i, \qquad \text{and} \qquad \phi_i = \frac{\pi}{2}. \tag{8.25}$$

With Assumption 8.1, the iterative solution (8.24) can be rewritten as

$$\begin{pmatrix} (U_1^T x)_{k,i} \\ (U_1^T v)_{k,i} \end{pmatrix} = \begin{pmatrix} \cos(kh\omega_i) & \frac{\sin(kh\omega_i)}{\omega_i} \\ -\omega_i \sin(kh\omega_i) & -\cos(kh\omega_i) \end{pmatrix} \begin{pmatrix} (U_1^T x)_{0,i} \\ (U_1^T v)_{0,i} \end{pmatrix}. \tag{8.26}$$

Theorem 8.2 *For every ill-conditioned eigen-direction, with every initial condition $(x_0, 0)^T$, if Algorithm 1 is implemented at $\|v_{iter}\| \leq \|v\|$, then there exists an eigenvalue ω_i^2 such that*

$$k\omega_i h \geq \frac{\pi}{2}.$$

Proof When $\|v_{iter}\| \leq \|v\|$, then $\left\|U_1^T v_{iter}\right\| \leq \left\|U_1^T v\right\|$. While for the $\left\|U_1^T v\right\|$, we can write in the analytical form

$$\left\|U_1^T v\right\| = \sqrt{\sum_{i=1}^{n} \omega_i^2 (U_1 x_0)_i^2 \sin^2(kh\omega_i)},$$

if there is no $k\omega_i h < \frac{\pi}{2}$, $\left\|U_1^T v\right\|$ increases with k increasing. □

For some i such that $k\omega_i h$ approximating $\frac{\pi}{2}$, we have

$$\begin{aligned} \frac{\left|(U_1^T x)_{k+1,i}\right|}{\left|(U_1^T x)_{k,i}\right|} &= \frac{\cos\left((k+1)h\omega_i\right)}{\cos\left(kh\omega_i\right)} \\ &= e^{\ln\cos((k+1)h\omega_i) - \ln\cos(kh\omega_i)} \\ &= e^{-\tan(\xi)h\omega_i} \end{aligned} \tag{8.27}$$

where $\xi \in (kh\omega_i, (k+1)h\omega_i)$. Hence, with ξ approximating $\frac{\pi}{2}$, $\left|(U_1^T x)_{k,i}\right|$ approximates 0 with the linear convergence, but the coefficient will also decay with the rate $e^{-\tan(\xi)h\omega_i}$ with $\xi \to \frac{\pi}{2}$. With the Laurent expansion for $\tan\xi$ at $\frac{\pi}{2}$, i.e.,

$$\tan\xi = -\frac{1}{\xi - \frac{\pi}{2}} + \frac{1}{3}\left(\xi - \frac{\pi}{2}\right) + \frac{1}{45}\left(\xi - \frac{\pi}{2}\right)^3 + \mathcal{O}\left(\left(\xi - \frac{\pi}{2}\right)^5\right),$$

the coefficient has the approximating formula

$$e^{-\tan(\xi)h\omega_i} \approx e^{\frac{h\omega_i}{\xi - \frac{\pi}{2}}} \leq \left(\frac{\pi}{2} - \xi\right)^n,$$

where n is an arbitrary large real number in $\mathbb{R}^+$ for $\xi \to \frac{\pi}{2}$.

8.4 Experimental Demonstration

In this section, we implement the artificially dissipating energy algorithm (Algorithm 1), energy conservation algorithm (Algorithm 2), and the combined algorithm (Algorithm 3) into high-dimensional data for comparison with gradient method, momentum method, and Nesterov's accelerated gradient method (Fig. 8.7).

8.4.1 *Strongly Convex Function*

Here, we investigate the artificially dissipating energy algorithm (Algorithm 1) for the strongly convex function for comparison with gradient method, momentum method, and Nesterov's accelerated gradient method (strongly convex case) by the quadratic function as below:

$$f(x) = \frac{1}{2}x^T Ax + b^T x, \tag{8.28}$$

where A is symmetric and positive-definite matrix. The two cases are shown as below:

(a) The generate matrix A is 500×500 random positive define matrix with eigenvalue from $1e-6$ to 1 with one defined eigenvalue $1e-6$. The generate vector b follows i.i.d. Gaussian distribution with mean 0 and variance 1.
(b) The generate matrix A is the notorious example in Nesterov's book [Nes13], i.e.,

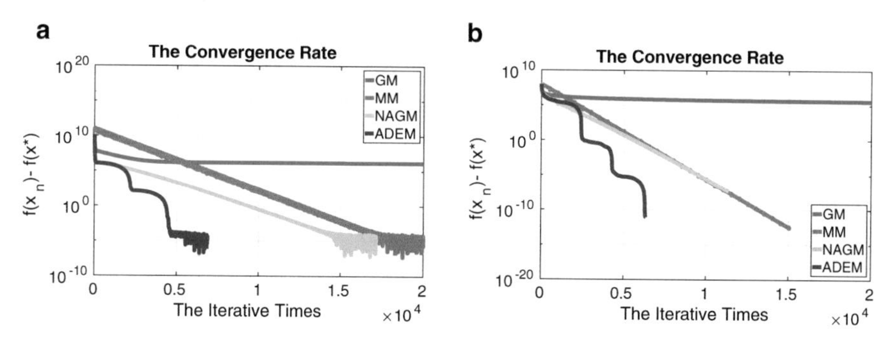

Fig. 8.7 Left: the case (**a**) with the initial point $x_0 = 0$. Right: the case (**b**) with the initial point $x_0 = 1000$

$$A = \begin{pmatrix} 2 & -1 & & & & \\ -1 & 2 & -1 & & & \\ & -1 & 2 & \ddots & & \\ & & \ddots & \ddots & \ddots & \\ & & & \ddots & \ddots & -1 \\ & & & & -1 & 2 \end{pmatrix},$$

the eigenvalues of the matrix are

$$\lambda_k = 2 - 2\cos\left(\frac{k\pi}{n+1}\right) = 4\sin^2\left(\frac{k\pi}{2(n+1)}\right),$$

and n is the dimension of the matrix A. The eigenvector can be solved by the second Chebyshev's polynomial. We implement $\dim(A) = 1000$ and b is zero vector. Hence, the smallest eigenvalue is approximating

$$\lambda_1 = 4\sin^2\left(\frac{\pi}{2(n+1)}\right) \approx \frac{\pi^2}{1001^2} \approx 10^{-5}.$$

8.4.2 *Non-Strongly Convex Function*

Here, we investigate the artificially dissipating energy algorithm (Algorithm 1) for the non-strongly convex function for comparison with gradient method, Nesterov's accelerated gradient method (non-strongly convex case) by the log-sum-exp function as below:

$$f(x) = \rho \log\left[\sum_{i=1}^{n} \exp\left(\frac{\langle a_i, x\rangle - b_i}{\rho}\right)\right], \tag{8.29}$$

where A is the $m \times n$ matrix with a_i, $(i = 1, \ldots, m)$ and the column vector of A and b is the $n \times 1$ vector with component b_i. ρ is the parameter. We show the experiment in (8.29): the matrix $A = \left(a_{ij}\right)_{m\times n}$ and the vector $b = (b_i)_{n\times 1}$ are set by the entry following i.i.d Gaussian distribution for the parameter $\rho = 5$ and $\rho = 10$ (Fig. 8.8).

8.4.3 *Non-Convex Function*

For the non-convex function, we exploit classical test function, known as artificial landscape, to evaluate characteristics of optimization algorithms from general performance and precision. In this paper, we show our algorithms implementing on

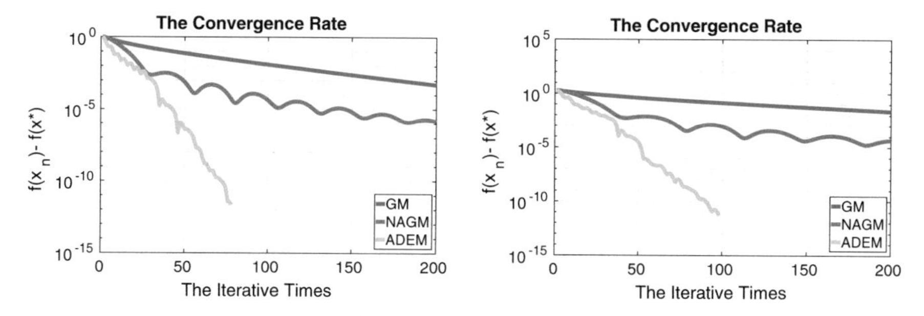

Fig. 8.8 The convergence rate is shown from the initial point $x_0 = 0$. Left: $\rho = 5$; Right: $\rho = 10$

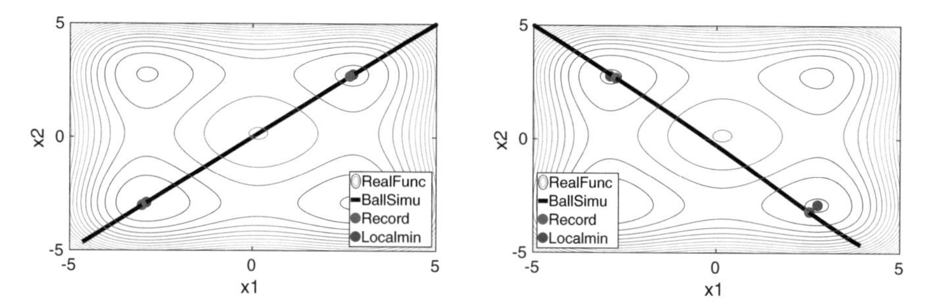

Fig. 8.9 Detecting the number of the local minima of 2-D Styblinski–Tang function by Algorithm 3 with step length $h = 0.01$. The red points are recorded by Algorithm 2 and the blue points are the local minima by Algorithm 1. Left: The initial position $(5, 5)$; Right: The initial position $(-5, 5)$

the Styblinski–Tang function and Shekel function, which is recorded in the virtual library of simulation experiments.[1] Firstly, we investigate Styblinski–Tang function, i.e.,

$$f(x) = \frac{1}{2}\sum_{i=1}^{d}\left(x_i^4 - 16x_i^2 + 5x_i\right), \tag{8.30}$$

to demonstrate the general performance of Algorithm 2 to track the number of local minima and then find the local minima by Algorithm 3 (Fig. 8.9).

To the essential 1-D non-convex Styblinski–Tang function of high dimension, we implement Algorithm 3 to obtain the precision of the global minima as below.

The global minima calculated at the position $(-2.9035, -2.9035, \ldots)$ is -391.6617 as shown in Table 8.1. And the real global minima at $(-2.903534, -2.903534, \ldots)$ is $-39.16599 \times 10 = -391.6599$.

[1]https://www.sfu.ca/~ssurjano/index.html.

Table 8.1 The example for ten-dimensional Styblinski–Tang function from two initial positions

	Local_min1	Local_min2	Local_min3	Local_min4
Initial position	$(5, 5, \ldots)$	$(5, 5, \ldots)$	$(5, -5, \ldots)$	$(5, -5, \ldots)$
Position	$(2.7486, 2.7486, \ldots)$	$(-2.9035, -2.9035, \ldots)$	$(2.7486, -2.9035, \ldots)$	$(-2.9035, 2.7486, \ldots)$
Function value	-250.2945	-391.6617	-320.9781	-320.9781

Furthermore, we demonstrate the numerical experiment from Styblinski–Tang function to more complex Shekel function

$$f(x) = -\sum_{i=1}^{m} \left(\sum_{j=1}^{4} \left(x_j - C_{ji} \right)^2 + \beta_i \right)^{-1}, \tag{8.31}$$

where

$$\beta = \frac{1}{10} (1, 2, 2, 4, 4, 6, 3, 7, 5, 5)^T$$

and

$$C = \begin{pmatrix} 4.0 & 1.0 & 8.0 & 6.0 & 3.0 & 2.0 & 5.0 & 8.0 & 6.0 & 7.0 \\ 4.0 & 1.0 & 8.0 & 6.0 & 7.0 & 9.0 & 3.0 & 1.0 & 2.0 & 3.6 \\ 4.0 & 1.0 & 8.0 & 6.0 & 3.0 & 2.0 & 5.0 & 8.0 & 6.0 & 7.0 \\ 4.0 & 1.0 & 8.0 & 6.0 & 7.0 & 9.0 & 3.0 & 1.0 & 2.0 & 3.6 \end{pmatrix}.$$

(1) Case $m = 5$, the global minima at $x^\star = (4, 4, 4, 4)$ is $f(x^\star) = -10.1532$.

(a) From the position (10, 10, 10, 10), the experimental result with the step length $h = 0.01$ and the iterative times 3000 is shown as below:
Detect Position (Algorithm 2)

$$\begin{pmatrix} 7.9879 & 6.0136 & 3.8525 & 6.2914 & 2.7818 \\ 7.9958 & 5.9553 & 3.9196 & 6.2432 & 6.7434 \\ 7.9879 & 6.0136 & 3.8525 & 6.2914 & 2.7818 \\ 7.9958 & 5.9553 & 3.9196 & 6.2432 & 6.7434 \end{pmatrix}$$

Detect value

$$\begin{pmatrix} -5.0932 & -2.6551 & -6.5387 & -1.6356 & -1.7262 \end{pmatrix}$$

Final position (Algorithm 1)

$$\begin{pmatrix} 7.9996 & 5.9987 & 4.0000 & 5.9987 & 3.0018 \\ 7.9996 & 6.0003 & 4.0001 & 6.0003 & 6.9983 \\ 7.9996 & 5.9987 & 4.0000 & 5.9987 & 3.0018 \\ 7.9996 & 6.0003 & 4.0001 & 6.0003 & 6.9983 \end{pmatrix}$$

Final value

$$\begin{pmatrix} -5.1008 & -2.6829 & -10.1532 & -2.6829 & -2.6305 \end{pmatrix}$$

(b) From the position (3, 3, 3, 3), the experimental result with the step length $h = 0.01$ and the iterative times 1000 is shown as below:
Detect Position (Algorithm 2)

$$\begin{pmatrix} 3.9957 & 6.0140 \\ 4.0052 & 6.0068 \\ 3.9957 & 6.0140 \\ 4.0052 & 6.0068 \end{pmatrix}$$

Detect value

$$\begin{pmatrix} -10.1443 & -2.6794 \end{pmatrix}$$

Final position (Algorithm 1)

$$\begin{pmatrix} 4.0000 & 5.9987 \\ 4.0001 & 6.0003 \\ 4.0000 & 5.9987 \\ 4.0001 & 6.0003 \end{pmatrix}$$

Final value

$$\begin{pmatrix} -10.1532 & -2.6829 \end{pmatrix}$$

(2) Case $m = 7$, the global minima at $x^\star = (4, 4, 4, 4)$ is $f(x^\star) = -10.4029$.

(a) From the position (10, 10, 10, 10), the experimental result with the step length $h = 0.01$ and the iterative times 3000 is shown as below:
Detect Position (Algorithm 2)

$$\begin{pmatrix} 7.9879 & 6.0372 & 3.1798 & 5.0430 & 6.2216 & 2.6956 \\ 8.0041 & 5.9065 & 3.8330 & 2.8743 & 6.2453 & 6.6837 \\ 7.9879 & 6.0372 & 3.1798 & 5.0430 & 6.2216 & 2.6956 \\ 8.0041 & 5.9065 & 3.8330 & 2.8743 & 6.2453 & 6.6837 \end{pmatrix}$$

Detect value

$$\begin{pmatrix} -5.1211 & -2.6312 & -0.9428 & -3.3093 & -1.8597 & -1.5108 \end{pmatrix}$$

Final position (Algorithm 1)

$$\begin{pmatrix} 7.9995 & 5.9981 & 4.0006 & 4.9945 & 5.9981 & 3.0006 \\ 7.9996 & 5.9993 & 3.9996 & 3.0064 & 5.9993 & 7.0008 \\ 7.9995 & 5.9981 & 4.0006 & 4.9945 & 5.9981 & 3.0006 \\ 7.9996 & 5.9993 & 3.9996 & 3.0064 & 5.9993 & 7.0008 \end{pmatrix}$$

Final value

$$\begin{pmatrix} -5.1288 & -2.7519 & -10.4029 & -3.7031 & -2.7519 & -2.7496 \end{pmatrix}$$

(b) From the position (3, 3, 3, 3), the experimental result with the step length $h = 0.01$ and the iterative times 1000 is shown as below:
Detect Position (Algorithm 2)

$$\begin{pmatrix} 4.0593 & 3.0228 \\ 3.9976 & 7.1782 \\ 4.0593 & 3.0228 \\ 3.9976 & 7.1782 \end{pmatrix}$$

Detect value

$$\begin{pmatrix} -9.7595 & -2.4073 \end{pmatrix}$$

Final position (Algorithm 1)

$$\begin{pmatrix} 4.0006 & 3.0006 \\ 3.9996 & 7.0008 \\ 4.0006 & 3.0006 \\ 3.9996 & 7.0008 \end{pmatrix}$$

Final value

$$\begin{pmatrix} -10.4029 & -2.7496 \end{pmatrix}$$

(3) Case $m = 10$, the global minima at $x^{\star} = (4, 4, 4, 4)$ is $f(x^{\star}) = -10.5364$.

(a) From the position (10, 10, 10, 10), the experimental result with the step length $h = 0.01$ and the iterative times 3000 is shown as below:
Detect Position (Algorithm 2)

$$\begin{pmatrix} 7.9977 & 5.9827 & 4.0225 & 2.7268 & 6.1849 & 6.2831 & 6.3929 \\ 7.9942 & 6.0007 & 3.8676 & 7.3588 & 6.0601 & 3.2421 & 1.9394 \\ 7.9977 & 5.9827 & 4.0225 & 2.7268 & 6.1849 & 6.2831 & 6.3929 \\ 7.9942 & 6.0007 & 3.8676 & 7.3588 & 6.0601 & 3.2421 & 1.9394 \end{pmatrix}$$

Detect value

$$\begin{pmatrix} -5.1741 & -2.8676 & -7.9230 & -1.5442 & -2.4650 & -1.3703 & -1.7895 \end{pmatrix}$$

Final position (Algorithm 1)

$$\begin{pmatrix} 7.9995 & 5.9990 & 4.0007 & 3.0009 & 5.9990 & 6.8999 & 5.9919 \\ 7.9994 & 5.9965 & 3.9995 & 7.0004 & 5.9965 & 3.4916 & 2.0224 \\ 7.9995 & 5.9990 & 4.0007 & 3.0009 & 5.9990 & 6.8999 & 5.9919 \\ 7.9994 & 5.9965 & 3.9995 & 7.0004 & 5.9965 & 3.4916 & 2.0224 \end{pmatrix}$$

Final value

$$\begin{pmatrix} -5.1756 & -2.8712 & -10.5364 & -2.7903 & -2.8712 & -2.3697 & -2.6085 \end{pmatrix}$$

(b) From the position (3, 3, 3, 3), the experimental result with the step length $h = 0.01$ and the iterative times 1000 is shown as below:
Detect Position (Algorithm 2)

$$\begin{pmatrix} 4.0812 & 3.0206 \\ 3.9794 & 7.0173 \\ 4.0812 & 3.0206 \\ 3.9794 & 7.0173 \end{pmatrix}$$

Detect value

$$\begin{pmatrix} -9.3348 & -2.7819 \end{pmatrix}$$

Final position (Algorithm 1)

$$\begin{pmatrix} 4.0007 & 3.0009 \\ 3.9995 & 7.0004 \\ 4.0007 & 3.0009 \\ 3.9995 & 7.0004 \end{pmatrix}$$

Final value

$$\begin{pmatrix} -10.5364 & -2.7903 \end{pmatrix}$$

8.5 Conclusion and Further Works

Based on the view for understanding arithmetical complexity from analytical complexity in the seminal book [Nes13] and the idea for viewing optimization from differential equation in the novel blog,[2] we propose some original algorithms based on Newton's second law with the kinetic energy observable and controllable in the computational process firstly. Although our algorithm cannot fully solve the global optimization problem, or it is dependent on the trajectory path, this work introduces the Hamilton system essential to optimization such that it is possible that the global minima can be obtained. Our algorithms are easy to implement and own a more rapid convergence rate.

For the theoretical view, the Hamilton system is closer to nature and a lot of fundamental work have appeared in the previous century, such as KAM theory, Nekhoroshev estimate, operator spectral theory, and so on. Are these beautiful and essentially original work used to understand and improve the algorithm for optimization and machine learning? Also, in establishing the convergence rate, the matrix containing the trigonometric function can be hard to estimate. Researchers have proposed some methods for estimating the trigonometric matrix based on spectral theory. For the numerical scheme, we only exploit the simple first-order symplectic Euler method. Several more efficient schemes, such as Störmer–Verlet scheme, symplectic Runge–Kutta scheme, order condition method, and so on, are proposed in [Nes13].

These schemes can make the algorithms in this paper more efficient and accurate. For the optimization, the method we proposed is only about unconstrained problem. In the nature, the classical Newton's second law, or the equivalent expression—Lagrange mechanics and Hamilton mechanics, is implemented on the manifold in the almost real physical world. In other words, a natural generalization is from unconstrained problem to constrained problem for our proposed algorithms. A more natural implementation is the geodesic descent in [LY$^+$84]. Similar to the development of the gradient method from smooth condition to nonsmooth condition, our algorithms can be generalized to nonsmooth condition by the subgradient. For application, we will implement our algorithms to non-negative matrix factorization, matrix completion, and deep neural network and speed up the training of the objective function. Meanwhile, we apply the algorithms proposed in this paper to the maximum likelihood estimator and maximum a posteriori estimator in statistics.

Starting from Newton's second law, we implement only a simple particle in classical mechanics, or macroscopic world. A natural generalization is from the macroscopic world to the microscopic world. In the field of fluid dynamics, the Newton's second law is expressed by Euler equation, or more complex Navier–Stokes equation. An important topic from fluid dynamics is geophysical fluid dynamics, containing atmospheric science and oceanography. Especially, a key

[2]http://www.offconvex.org/2015/12/11/mission-statement/.

feature in the oceanography different from atmospheric science is the topography, which influences mainly vector field of the fluid. So many results have been demonstrated based on many numerical modeling, such as the classical POM,[3] HYCOM,[4] ROMS,[5] and FVCOM.[6] A reverse idea is that if we view the potential function in black box is the topography, we observe the changing of the fluid vector field to find the number of local minima in order to obtain the global minima with a suitable initial vector field. A more adventurous idea is to generalize the classical particle to the quantum particle. For quantum particle, the Newton's second law is expressed by the energy form that is from the view of Hamilton mechanics, which is the starting point for the proposed algorithm in this paper. The particle appears in the wave form in a microscopic world. When the wave meets the potential barrier, the tunneling phenomena will appear. The tunneling phenomena will also appear in higher dimensions. It is very easy to observe the tunneling phenomena in the physical world. However, if we attempt to compute this phenomena, the problem becomes NP-hard. Only if quantum computing is used is the phenomena very easy to simulate, as we can find the global minima by binary section search. That is, if there exist tunneling phenomena in the upper level, the algorithm will continue to detect this in the upper level, otherwise go to the lower level. In the quantum world, it needs only $\mathcal{O}(\log n)$ times to find the global minima rather than becoming NP-hard.

[3] http://ofs.dmcr.go.th/thailand/model.html.

[4] https://hycom.org/.

[5] https://www.myroms.org/.

[6] http://fvcom.smast.umassd.edu/.

Part III
Mathematical Framework for Machine Learning: Application Part

Chapter 9
Improved Sample Complexity in Sparse Subspace Clustering with Noisy and Missing Observations

In this chapter, we show the results of the new CoCoSSC algorithm. The content is organized as follows: The main results concerning CoCoSSC algorithm are shown in Sect. 9.1. Following Sect. 9.1, we show the full proofs in Sect. 9.2. In Sect. 9.3, we show the performance for CoCoSSC algorithm and some related algorithms numerically. Finally, we conclude this work with some future directions.

9.1 Main Results About CoCoSSC Algorithm

We introduce our main results by analyzing the performance of CoCoSSC under both the Gaussian noise model and the missing data model. Similar to [WX16], the quality of the computed self-similarity matrix $\{\boldsymbol{c}_i\}_{i=1}^N$ is assessed using a *subspace detection property (SDP)*:

Definition 9.1 (Subspace Detection Property (SDP), [WX16]) The self-similarity matrix $\{\boldsymbol{c}_i\}_{i=1}^N$ satisfies the *subspace detection property* if (1) for every $i \in [N]$, $\boldsymbol{c}_i$ is a non-zero vector; and (2) for every $i, j \in [N]$, $\boldsymbol{c}_{ij} \neq 0$ implies that $\boldsymbol{x}_i$ and $\boldsymbol{x}_j$ belong to the same cluster.

Intuitively, the subspace detection property asserts that the self-similarity matrix $\{\boldsymbol{c}_i\}_{i=1}^N$ has *no false positives*, where every non-zero entry in $\{\boldsymbol{c}_i\}_{i=1}^n$ links two data points $\boldsymbol{x}_i$ and $\boldsymbol{x}_j$ to the same cluster. The first property in Definition 9.1 further rules out the trivial solution of $\boldsymbol{c}_i \equiv \boldsymbol{0}$.

Part of this chapter is in the paper titled "Improved Sample Complexity in Sparse Subspace Clustering with Noisy and Missing Observations" by Yining Wang, Bin Shi et al. (2018) presently under review for publication in AISTATS.

B. Shi, S. S. Iyengar, *Mathematical Theories of Machine Learning - Theory and Applications*, https://doi.org/10.1007/978-3-030-17076-9_9

The SDP stated in Definition 9.1 is, however, *not* sufficient for the success of a follow-up spectral clustering algorithm, or any clustering algorithm, as the "similarity graph" constructed by connecting every pairs of (i, j) with $c_{ij} \neq 0$ might be poorly connected. Such "graph connectivity" is a well-known open problem in sparse subspace clustering [NH11] and remains largely unsolved except under strong assumptions [WWS16]. Nevertheless, in practical scenarios the SDP criterion correlates reasonably well with clustering performance [WX16, WWS15a] and therefore we choose to focus on the SDP success condition only.

9.1.1 The Non-Uniform Semi-Random Model

We adopt the following non-uniform semi-random model throughout the paper:

Definition 9.2 (Non-Uniform Semi-Random Model) Suppose $\boldsymbol{y}_i$ belongs to cluster $\mathcal{S}_\ell$ and let $\boldsymbol{y}_i = \mathbf{U}_\ell \boldsymbol{\alpha}_i$, where $\mathbf{U}_\ell \in \mathbb{R}^{n \times d_\ell}$ is an orthonormal basis of $\mathcal{U}_\ell$ and $\boldsymbol{\alpha}_i$ is a d_ℓ-dimensional vector with $\|\boldsymbol{\alpha}_i\|_2 = 1$. We assume that $\boldsymbol{\alpha}_i$ are i.i.d. distributed according to an unknown underlying distribution P_ℓ, and that the density p_ℓ associated with P_ℓ satisfies

$$0 < \underline{C} \cdot p_0 \leq p_\ell(\boldsymbol{\alpha}) \leq \overline{C} \cdot p_0 < \infty \qquad \forall \boldsymbol{\alpha} \in \mathbb{R}^{d_\ell}, \quad \|\boldsymbol{\alpha}\|_2 = 1$$

for some constants $\underline{C}, \overline{C}$, where p_0 is the density of the uniform measure on $\{\boldsymbol{u} \in \mathbb{R}^{d_\ell} : \|\boldsymbol{u}\|_2 = 1\}$.

Remark 9.1 Our non-uniform semi-random model ensures that $\|\boldsymbol{y}_i\|_2 = 1$ for all $i \in [N]$, a common normalizing assumption made in previous works on sparse subspace clustering [SC12, SEC14, WX16]. However, such a property is only used in our theoretical analysis, and in our CoCoLasso algorithm the norms of $\{y_i\}_{i=1}^N$ are assumed unknown. Indeed, if the exact norms of $\|\boldsymbol{y}_i\|_2$ are known to the data analyst the sample complexity in our analysis can be further improved, as we remarked in Remark 9.3.

The non-uniform semi-random model considers fixed (deterministic) subspaces $\{\mathcal{S}_\ell\}$, but assumes that data points within each low-dimensional subspace are independently generated from an unknown distribution P_ℓ with densities bounded away and above from below. This helps simplifying the "inter-subspace incoherence" (Definition 9.6) in our proof and yields interpretable results.

Compared with existing definitions of semi-random models [SC12, WX16, HB15, PCS14], the key difference is that in our model data are *not* uniformly distributed on each low-dimensional subspace. Instead, it is assumed that the data points are i.i.d., and that the data density is bounded away from both above and below. Such non-uniformity rules out algorithms that exploit the $\mathbb{E}[\boldsymbol{y}_i] = \mathbf{0}$ property in traditional semi-random models which is too strong and rarely holds true in practice.

Because the underlying subspaces are fixed, quantities that characterize the "affinity" between these subspace are needed because closer subspaces are harder to distinguish from each other. We adopt the following affinity measure, which was commonly used in previous works on sparse subspace clustering [WX16, WWS15a, CJW17]:

Definition 9.3 (Subspace Affinity) Let $\mathcal{U}_j$ and $\mathcal{U}_k$ be two linear subspaces of $\mathbb{R}^n$ of dimension d_j and d_k. The *affinity* between $\mathcal{U}_j$ and $\mathcal{U}_k$ is defined as $\chi_{j,k}^2 := \cos^2\theta_{jk}^{(1)} + \cdots + \cos^2\theta_{jk}^{(\min(d_j,d_k))}$, where $\theta_{jk}^{(\ell)}$ is the ℓth canonical angle between $\mathcal{U}_j$ and $\mathcal{U}_k$.

Remark 9.2 $\chi_{jk} = \|\mathbf{U}_j^\top \mathbf{U}_k\|_F$, where $\mathbf{U}_j \in \mathbb{R}^{n\times d_j}, \mathbf{U}_k \in \mathbb{R}^{n\times d_k}$ are orthonormal basis of $\mathcal{U}_j, \mathcal{U}_k$.

Throughout the paper we also write $\chi := \max_{j\neq k}\chi_{j,k}$.

For the missing data model, we need the following additional "inner-subspace" incoherence of the subspaces to ensure that the observed data entries contain sufficient amount of information. Such incoherence assumptions were widely adopted in the matrix completion community [CR09, KMO10, Rec11].

Definition 9.4 (Inner-Subspace Incoherence) Fix $\ell \in [L]$ and let $\mathbf{U}_\ell \in \mathbb{R}^{n\times d_\ell}$ be an orthonormal basis of subspace $\mathcal{U}_\ell$. The *subspace incoherence* of $\mathcal{U}_\ell$ is the smallest μ_ℓ such that

$$\max_{1\leq i\leq n} \|\boldsymbol{e}_i^\top \mathbf{U}_\ell\|_2^2 \leq \mu_\ell d_\ell/n.$$

With the above definitions, we are now ready to state the following two theorems which give sufficient success conditions for the self-similarity matrix $\{\boldsymbol{c}_i\}_{i=1}^n$ produced by CoCoLasso.

Theorem 9.1 (The Gaussian Noise Model) *Suppose* $\lambda \asymp 1/\sqrt{d}$ *and* $\boldsymbol{\Delta}_{jk} \asymp \sigma^2\sqrt{\frac{\log N}{n}}$ *for all* $j,k \in [N]$. *Suppose also that* $N_\ell \geq 2\overline{C}d_\ell/\underline{C}$. *There exists a constant* $K_0 > 0$ *such that, if*

$$\sigma < K_0\left(n/d^3\log^2(\overline{C}N/\underline{C})\right)^{1/4},$$

then the optimal solution $\{\boldsymbol{c}_i\}_{i=1}^N$ *of the* CoCoSSC *estimator satisfies the subspace detection property (SDP) with probability* $1 - O(N^{-10})$.

Theorem 9.2 (The Missing Data Model) *Suppose* $\lambda \asymp 1/\sqrt{d}$, $\boldsymbol{\Delta}_{jk} \asymp \frac{\mu d\log N}{\rho\sqrt{n}}$ *for* $j \neq k$ *and* $\boldsymbol{\Delta}_{jk} \asymp \frac{\mu d\log N}{\rho^{3/2}\sqrt{n}}$ *for* $j = k$. *Suppose also that* $N_\ell \geq 2\overline{C}d_\ell/\underline{C}$. *There exists a constant* $K_1 > 0$ *such that, if*

$$\rho > K_1 \max\left\{(\mu\chi d^{5/2}\log^2 N)^{2/3}\cdot n^{-1/3}, (\mu^2 d^{7/2}\log^2 N)^{2/5}\cdot n^{-2/5}\right\},$$

then the optimal solution $\{\boldsymbol{c}_i\}_{i=1}^N$ *of the* CoCoSSC *estimator satisfies the subspace detection property (SDP) with probability* $1 - O(N^{-10})$.

Remark 9.3 If the norms of the data points $\|\boldsymbol{y}_i\|_2$ are exactly known and can be explicitly used in algorithm design, the diagonal terms of $\mathbf{A}$ in Eq. (4.1) can be directly set to $\mathbf{A}_{ii} = \|\boldsymbol{y}_i\|_2^2$ in order to avoid the ψ_2 concentration term in our proof (Definition 9.5). This would improve the sample complexity in the success condition to $\rho > \Omega(n^{-1/2})$, matching the sample complexity in linear regression problems with missing design entries [WWBS17].

Theorems 9.1 and 9.2 show that when the noise magnitude (σ in the Gaussian noise model and ρ^{-1} in the missing data model) is sufficiently small, a careful choice of tuning parameter λ results in a self-similarity matrix $\{\boldsymbol{c}_i\}$ satisfying the subspace detection property. Furthermore, the maximum amount of noise our method can tolerate is $\sigma = O(n^{1/4})$ and $\rho = \Omega(\chi^{2/3}n^{-1/3} + n^{-2/5})$, which improves over the sample complexity of existing methods (see Table 4.1).

9.1.2 *The Fully Random Model*

When the underlying subspaces $\mathcal{U}_1, \cdots, \mathcal{U}_L$ are independently uniformly sampled, a model referred to as the *fully random* model in the literature [SC12, SEC14, WX16], the success condition in Theorem 9.2 can be further simplified:

Corollary 9.1 *Suppose subspaces* $\mathcal{U}_1, \cdots, \mathcal{U}_L$ *have the same intrinsic dimension* d *and are uniformly sampled, the condition in Theorem 9.2 can be simplified to*

$$\rho > \widetilde{K}_1(\mu^2 d^{7/2} \log^2 N)^{2/5} \cdot n^{-2/5},$$

where $\widetilde{K}_1 > 0$ *is a new universal constant.*

Corollary 9.1 shows that in the fully random model, the $\chi^{2/3}n^{-1/3}$ term in Theorem 9.2 is negligible and the success condition becomes $\rho = \Omega(n^{-2/5})$, strictly improving existing results (see Table 4.1).

9.2 Proofs

In this section we give proofs of our main results. Due to space constraints, we only give a proof framework and leave the complete proofs of all technical lemmas to the appendix.

9.2.1 Noise Characterization and Feasibility of Pre-Processing

Definition 9.5 (Characterization of Noise Variables) $\{z_i\}$ are independent random variables and $\mathbb{E}[z_i] = \mathbf{0}$. Furthermore, there exist parameters $\psi_1, \psi_2 > 0$ such that with probability $1 - O(N^{-10})$ the following holds uniformly for all $i, j \in [N]$:

$$\left|z_i^\top \boldsymbol{y}_j\right| \le \psi_1\sqrt{\frac{\log N}{n}}; \qquad \left|z_i^\top z_j - \mathbb{E}[z_i^\top z_j]\right| \le \begin{cases} \psi_1\sqrt{\frac{\log N}{n}} & i \neq j; \\ \psi_2\sqrt{\frac{\log N}{n}} & i = j. \end{cases}$$

Proposition 9.1 *Suppose* $\boldsymbol{\Delta}$ *are set as* $\boldsymbol{\Delta}_{jk} \ge 3\psi_1\sqrt{\frac{\log N}{n}}$ *for* $j \neq k$ *and* $\boldsymbol{\Delta}_{jk} \ge 3\psi_2\sqrt{\frac{\log N}{n}}$ *for* $j = k$. *Then with probability* $1 - O(N^{-10})$ *the set* S *defined in Eq. (4.1) is not empty.*

The following two lemmas derive explicit bounds on ψ_1 and ψ_2 for the two noise models.

Lemma 9.1 *The Gaussian noise model satisfies Definition 9.5 with* $\psi_1 \lesssim \sigma^2$ *and* $\psi_2 \lesssim \sigma^2$.

Lemma 9.2 *Suppose* $\rho = \Omega(n^{-1/2})$. *The missing data model satisfies Definition 9.5 with* $\psi_1 \lesssim \rho^{-1}\mu d\sqrt{\log N}$ *and* $\psi_2 \lesssim \rho^{-3/2}\mu d\sqrt{\log N}$, *where* $d = \max_{\ell\in[L]} d_\ell$ *and* $\mu = \max_{\ell\in[L]} \mu_\ell$.

9.2.2 Optimality Condition and Dual Certificates

We first write down the dual problem of CoCoSSC:

$$\text{Dual CoCoSSC}: \qquad \boldsymbol{v}_i = \arg\max_{\boldsymbol{v}_i\in\mathbb{R}^N} \tilde{\boldsymbol{x}}_i^\top \boldsymbol{v}_i - \frac{1}{2\lambda}\|\boldsymbol{v}_i\|_2^2 \quad s.t. \quad \left\|\widetilde{\mathbf{X}}_{-i}^\top \boldsymbol{v}_i\right\|_\infty \le 1. \tag{9.1}$$

Lemma 9.3 (Dual Certificate, Lemma 12 of [WX16]) *Suppose there exists triplet* $(\boldsymbol{c}, \boldsymbol{e}, \boldsymbol{v})$ *such that* $\tilde{\boldsymbol{x}}_i = \widetilde{\mathbf{X}}_{-i}\boldsymbol{c} + \boldsymbol{e}$, $\boldsymbol{c}$ *has support* $S \subseteq T \subseteq [N]$, *and that* $\boldsymbol{v}$ *satisfies*

$$[\widetilde{\mathbf{X}}_{-i}]_S^\top \boldsymbol{v} = \mathrm{sgn}(\boldsymbol{c}_S), \quad \boldsymbol{v} = \lambda\boldsymbol{e}, \quad \left\|[\widetilde{\mathbf{X}}_{-i}]_{T\cap S^c}^\top \boldsymbol{v}\right\|_\infty \le 1, \quad \left\|[\widetilde{\mathbf{X}}_{-i}]_{T^c}^\top \boldsymbol{v}\right\|_\infty < 1,$$

then any optimal solution $\boldsymbol{c}_i$ *to Eq. (4.2) satisfies* $[\boldsymbol{c}_i]_{T^c} = \mathbf{0}$.

To construct such a dual certificate and to de-couple potential statistical dependency, we follow [WX16] to consider a constrained version of the optimization problem. Let $\widetilde{\mathbf{X}}_{-i}^{(\ell)}$ denote the data matrix of all but $\tilde{\boldsymbol{x}}_i$ in cluster $\mathcal{S}_\ell$. The constrained problems are defined as follows:

$$\text{Constrained Primal :} \quad \tilde{c}_i = \arg\min_{c_i \in \mathbb{R}^{N_\ell - 1}} \|c_i\|_1 + \lambda/2 \cdot \|\tilde{x}_i - \widetilde{\mathbf{X}}_{-i}^{(\ell)} c_i\|_2^2; \tag{9.2}$$

$$\text{Constrained Dual :} \quad \tilde{v}_i = \arg\max_{v_i \in \mathbb{R}^{N_\ell - 1}} \tilde{x}_i^\top v_i - 1/(2\lambda) \cdot \|v_i\|_2^2 \quad s.t. \quad \|(\widetilde{\mathbf{X}}_{-i}^{(\ell)})^\top v_i\|_\infty \le 1. \tag{9.3}$$

With $c = [\tilde{c}_i, \mathbf{0}_{\mathcal{S}_{-\ell}}]$, $v = [\tilde{v}_i, \mathbf{0}_{\mathcal{S}_{-\ell}}]$, and $e = \tilde{x}_i - \widetilde{\mathbf{X}}_{-i}^{(\ell)} \tilde{c}_i$, the certificate satisfies the first three conditions in Lemma 9.3 with $T = \mathcal{S}_\ell$ and $S = \text{supp}(\tilde{c}_i)$. Therefore, we only need to establish that $|\langle \tilde{x}_j, \tilde{v}_i \rangle| < 1$ for all $\tilde{x}_j \notin \mathcal{S}_\ell$ to show no false discoveries, which we prove in the next section.

9.2.3 Deterministic Success Conditions

Define the following deterministic quantities as *inter-subspace incoherence* and *in-radius*, which are important quantities in deterministic analysis of sparse subspace clustering methods [SC12, WX16, SEC14].

Definition 9.6 (Inter-Subspace Incoherence) The inter-subspace incoherence $\tilde{\mu}$ is defined as $\tilde{\mu} := \max_{\ell \in [L]} \max_{y_i \in \mathcal{S}_\ell} \max_{y_j \notin \mathcal{S}_\ell} \left| \langle y_i, y_j \rangle \right|$.

Definition 9.7 (In-Radius) Define r_i as the radius of the largest ball inscribed in the convex body of $\{\pm \mathbf{Y}_{-j}^{(\ell)}\}$. Also define that $r := \min_{1 \le i \le N} r_i$.

The following lemma derives an upper bound on $|\langle \tilde{x}_j, \tilde{v}_i \rangle|$, which is proved in the appendix.

Lemma 9.4 *For every (i, j) belonging to different clusters, $|\langle \tilde{x}_j, \tilde{v}_i \rangle| \lesssim \lambda(1 + \|\tilde{c}_i\|_1)(\tilde{\mu} + \psi_1\sqrt{\log N/n})$, where $\|\tilde{c}_i\|_1 \lesssim r^{-1}(1 + r^{-1}\lambda(\psi_1 + \psi_2)\sqrt{\log N/n})$.*

Lemmas 9.3 and 9.4 immediately yield the following theorem:

Theorem 9.3 (No False Discoveries) *There exists an absolute constant $\kappa_1 > 0$ such that if*

$$\frac{\lambda}{r}\left(1 + \frac{\lambda}{r}(\psi_1 + \psi_2)\sqrt{\frac{\log N}{n}}\right) \cdot \left(\tilde{\mu} + \psi_1\sqrt{\frac{\log N}{n}}\right) < \kappa_1, \tag{9.4}$$

then the optimal solution c_i of the CoCoSSC *estimator in Eq. (4.2) has no false discoveries, that is, $c_{ij} = 0$ for all x_j that belongs to a different cluster of x_i.*

The following theorem shows conditions under which c_i is not the trivial solution $c_i = \mathbf{0}$.

Theorem 9.4 (Avoiding Trivial Solutions) *There exists an absolute constant $\kappa_2 > 0$ such that, if*

$$\lambda\left(r - \psi_1\sqrt{\frac{\log N}{n}}\right) > \kappa_2, \tag{9.5}$$

then the optimal solution $\mathbf{c}_i$ of the CoCoSSC *estimator in Eq. (4.2) is non-trivial, that is, $\mathbf{c}_i \neq \mathbf{0}$.*

Finally, we remark that choosing $r = c/\lambda$ for some small constant $c > 0$ (depending only on κ_1 and κ_2), the choice of λ satisfies both theorems 9.3 and 9.4 provided that

$$\max\left\{\frac{\psi_1}{r}\sqrt{\frac{\log N}{n}}, \frac{\tilde{\mu}}{r^2}, \frac{\tilde{\mu}(\psi_1+\psi_2)}{r^3}\sqrt{\frac{\log N}{n}}, \frac{\psi_1(\psi_1+\psi_2)}{r^3}\frac{\log N}{n}\right\} < \kappa_3 \tag{9.6}$$

for some sufficiently small absolute constant $\kappa_3 > 0$ that depends on κ_1, κ_2, and c.

9.2.4 *Bounding $\tilde{\mu}$ and r in Randomized Models*

Lemma 9.5 *Suppose $N_\ell = \Omega(\overline{C}d_\ell/\underline{C}_\ell)$. Under the non-uniform semi-random model, with probability $1 - O(N^{-10})$ it holds that $\tilde{\mu} \lesssim \chi\sqrt{\log(\overline{C}N/\underline{C})}$ and $r \gtrsim 1/\sqrt{d}$.*

Lemma 9.6 *Suppose $\mathcal{U}_1, \ldots, \mathcal{U}_L$ are independently uniformly sampled linear subspaces of dimension d in $\mathbb{R}^n$. Then with probability $1 - O(N^{-10})$ we have that $\chi \lesssim d\sqrt{\log N/n}$ and $\mu \lesssim \sqrt{\log N}$.*

9.3 Numerical Results

Experimental Settings and Methods We conduct numerical experiments based on synthetic generated data, using a computer with Intel Core i7 CPU (4 GHz) and 16 GB memory. Each synthetic data set has ambient dimension $n = 100$, intrinsic dimension $d = 4$, number of underlying subspaces $L = 10$, and a total number of $N = 1000$ unlabeled data points. The observation rate ρ and Gaussian noise magnitude σ vary in our simulations. Underlying subspaces are generated uniformly at random, corresponding to our fully random model. Each data point has an equal probability of being assigned to any cluster and is generated uniformly at random on its corresponding low-dimensional subspace.

We compare the performance (explained later) of our CoCoSSC approach, and two popular existing methods Lasso SSC and the de-biased Dantzig selector. The ℓ_1 regularized self-regression steps in both CoCoSSC and Lasso SSC are implemented using ADMM. The pre-processing step of CoCoSSC is implemented using alternating projections initialized at $\tilde{\mathbf{\Sigma}} = \mathbf{X}^\top\mathbf{X} - \mathbf{D}$. Unlike the theoretical recommendations, we choose $\mathbf{\Delta}$ in Eq. (4.1) to be very large (3×10^3 for diagonal entries and 10^3 for off-diagonal entries) for fast convergence. The de-biased Dantzig selector is implemented using linear programming.

Evaluation Measure We consider two measures to evaluate the performance of algorithms being compared. The first one evaluates the quality of the similarity matrix $\{\boldsymbol{c}_i\}_{i=1}^N$ by measuring how far (relatively) it deviates from having the subspace detection property. In particular, we consider the RelViolation metric propositioned in [WX16] defined as

$$\text{RelViolation}(C, \mathcal{M}) = (\textstyle\sum_{(i,j)\notin\mathcal{M}}|C|_{i,j})/(\sum_{(i,j)\in\mathcal{M}}|C|_{i,j}), \tag{9.7}$$

where $\mathcal{M}$ is the mask of ground truth with all (i, j) satisfying $\boldsymbol{x}_i, \boldsymbol{x}_j \in \mathcal{S}^{(\ell)}$ for some ℓ. A high RelViolation indicates frequent deviation from the subspace detection propositionerty and therefore poorer quality of $\{\boldsymbol{c}_i\}_{i=1}^N$.

For clustering results, we use the Fowlkes–Mallows index [FM83] to evaluate their quality. Suppose $\mathcal{A} \subseteq \{(i, j) \in [N] \times [N]\}$ consists of pairs of data points that are clustered together by a clustering algorithm, and $\mathcal{A}_0$ is the ground truth clustering. Define $TP = |\mathcal{A} \cap \mathcal{A}_0|$, $FP = |\mathcal{A} \cap \mathcal{A}_0^c|$, $FN = |\mathcal{A}^c \cap \mathcal{A}_0|$, $TN = |\mathcal{A}^c \cap \mathcal{A}_0^c|$. The Fowlkes–Mallows (FM) index is then expressed as

$$FM = \sqrt{TP^2/(TP + FP)(TP + FN)}.$$

The FM index of any two clusterings $\mathcal{A}$ and $\mathcal{A}_0$ is always between 0 and 1, with an FM index of one indicating perfectly identical clusterings and an FM index close to zero otherwise.

Results We first give a qualitative illustration of similarity matrices $\{\boldsymbol{c}_i\}_{i=1}^N$ produced by the three algorithms of Lasso SSC, de-biased Dantzig selector, and CoCoSSC in Fig. 9.1. We observe that the similarity matrix of Lasso SSC has several spurious connections, and both Lasso SSC and the de-biased Dantzig selector suffer from graph connectivity issues as signals within each block (cluster) are not very strong. On the other hand, the similarity matrix of CoCoSSC produces convincing signals within each block (cluster). This shows that our propositioned CoCoSSC approach not only has few false discoveries as predicted by our theoretical results, but also has much better graph connectivity which our theory did not attempt to cover.

In Fig. 9.2 we report the Fowlkes–Mallows (FM) index for clustering results and RelViolation scores of similarity matrices $\{\boldsymbol{c}_i\}_{i=1}^N$ under various noise magnitude (σ) and observation rates (ρ) settings. A grid of tuning parameter values λ are attempted and the one leading to the best performance is reported. It is observed that our propositioned CoCoLasso consistently outperforms its competitors Lasso SSC

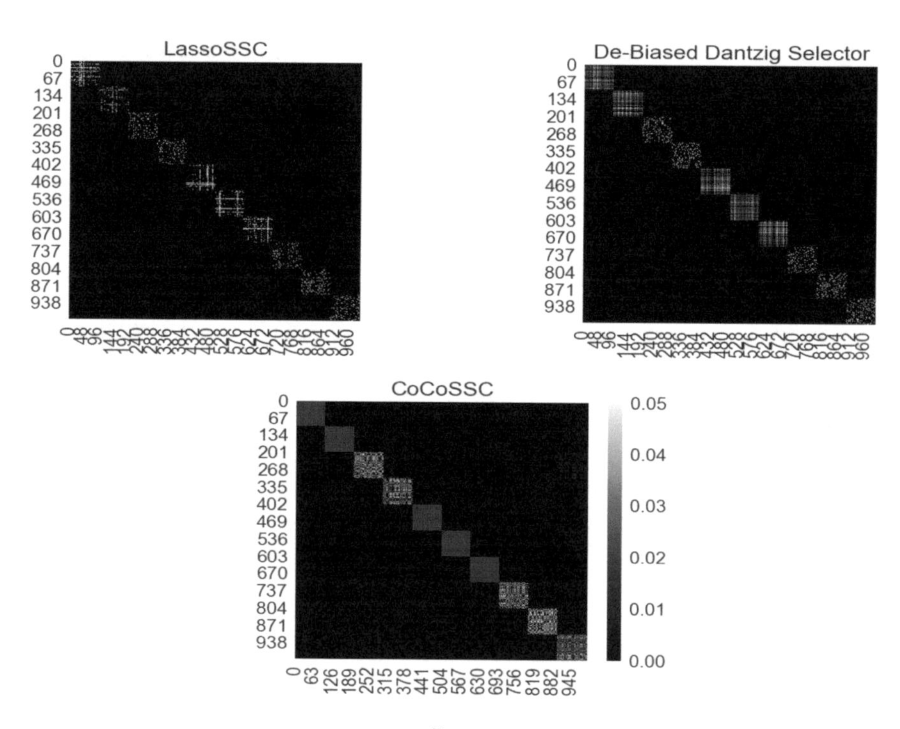

Fig. 9.1 Heatmaps of similarity matrices $\{c_i\}_{i=1}^N$, with brighter colors indicating larger absolute values of matrix entries. Left: LassoSSC; Middle: De-biased Dantzig selector; Right: CoCoSSC

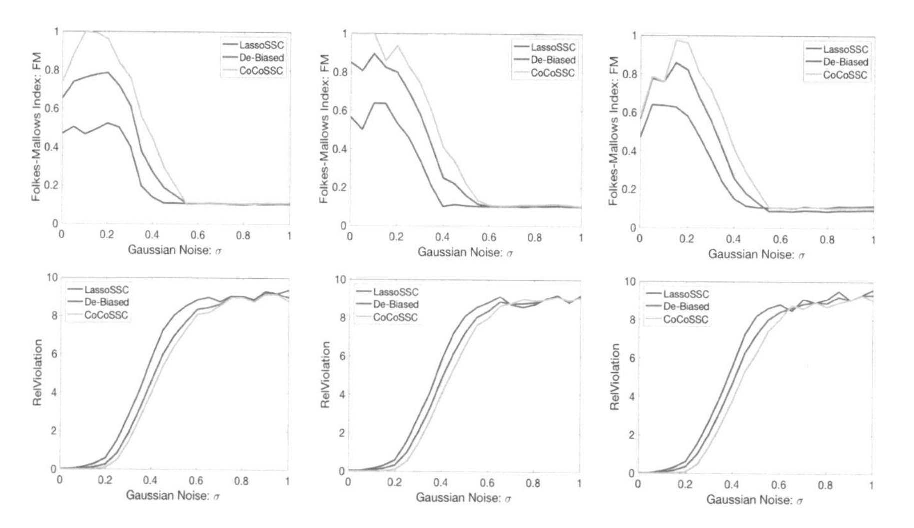

Fig. 9.2 The Fowlkes–Mallows (FM) index of clustering results (top row) and RelViolation scores (bottom row) of the three methods, with noise of magnitude σ varying from 0 to 1. Left column: missing rate $1 - \rho = 0.03$, middle column: $1 - \rho = 0.25$, right column: $1 - \rho = 0.9$

and de-biased Dantzig selector. Furthermore, CoCoLasso is very computationally efficient and converges in 8–15 seconds on each synthetic data set. On the other hand, de-biased Dantzig selector is computationally very expensive and typically takes over 100 seconds to converge.

9.4 Technical Details

Proof of Proposition 9.1 By Definition 9.5 we know that $|\widetilde{\boldsymbol{\Sigma}}_{-i} - \mathbf{Y}_{-i}^T\mathbf{Y}_{-i}| \le |\boldsymbol{\Delta}|$ in an element-wise sense. Also note that $\mathbf{Y}^\top\mathbf{Y}$ is positive semidefinite. Thus, $\mathbf{Y}^\top\mathbf{Y} \in S$. □

Proofs of Lemmas 9.1 and 9.2 Lemma 9.1 is proved in [WX16]. See Lemmas 17 and 18 of [WX16] and note that $\mathbb{E}[z_i^\top z_i] = \sigma^2$.

We next prove Lemma 9.2. We first consider $|z_i^\top \boldsymbol{y}_j|$. Let $z = z_i$, $\boldsymbol{y} = \boldsymbol{y}_i$, $\tilde{\boldsymbol{y}} = \boldsymbol{y}_j$, and $\boldsymbol{r} = R_{j\cdot}$. Define $T_i := z_i \boldsymbol{y}_i = (1 - \boldsymbol{r}_i/\rho)\boldsymbol{y}_i\tilde{\boldsymbol{y}}_j$. Because $\boldsymbol{r}$ is independent of $\boldsymbol{y}$ and $\tilde{\boldsymbol{y}}$, we have that $\mathbb{E}[T_i] = 0$, $\mathbb{E}[T_i^2] \le \boldsymbol{y}_i^2\tilde{\boldsymbol{y}}_i{}^2/\rho \le \mu^2 d^2/\rho n^2$, and $|T_i| \le \mu d/\rho n =: M$ almost surely. Using Bernstein's inequality, we know that with probability $1 - O(N^{-10})$

$$|z_i^\top \boldsymbol{y}_j| = \left|\sum_{i=1}^{T} T_i\right| \lesssim \sqrt{\sum_{i=1}^{n} \mathbb{E}[T_i^2]\cdot \log N} + M\log N \lesssim \mu d\sqrt{\frac{\log^2 N}{\rho n}}.$$

We next consider $|z_i^\top z_j|$ and the $i \neq j$ case. Let $\boldsymbol{y} = \boldsymbol{y}_i$, $\tilde{\boldsymbol{y}} = \boldsymbol{y}_j$, $\boldsymbol{r} = R_{i\cdot}$, and $\tilde{\boldsymbol{r}} = R_{j\cdot}$. By definition of μ, we have that $\|\boldsymbol{y}\|_\infty^2 \le \mu d_i/n$ and $\|\tilde{\boldsymbol{y}}\|_\infty^2 \le \mu d_j/n$. Define $T_i := z_i\tilde{z}_i = (1 - \boldsymbol{r}_i/\rho)(1 - \tilde{\boldsymbol{r}}_i/\rho)\cdot \boldsymbol{y}_i\tilde{\boldsymbol{y}}_i$. Because $\boldsymbol{r}$ and $\tilde{\boldsymbol{r}}$ are independent, $\mathbb{E}[T_i] = 0$, $\mathbb{E}[T_i^2] \le \boldsymbol{y}_i^2\tilde{\boldsymbol{y}}_i^2/\rho^2 \le \mu^2 d^2/\rho^2 n^2$, and $|T_i| \le \mu d/\rho^2 n =: M$ almost surely. Using Bernstein's inequality, we know that with probability $1 - O(N^{-10})$

$$\left|\sum_{i=1}^{n} T_i\right| \lesssim \sqrt{\sum_{i=1}^{n} \mathbb{E}[T_i^2]\cdot \log N} + M\log N \lesssim \frac{\mu d}{\rho}\sqrt{\frac{\log^2 N}{n}},$$

where the last inequality holds because $\rho = O(n^{-1/2})$.

Finally is the case of $|z_i^\top z_j|$ and $i = j$. Let again $z := z_i = z_j$. Define $T_i := z_i^2 - \mathbb{E}[z_i^2] = (1 - \boldsymbol{r}_i/\rho)^2\boldsymbol{y}_i^2 - (1-\rho)^2/\rho \cdot \boldsymbol{y}_i^2$. It is easy to verify that $\mathbb{E}[T_i] = 0$, $\mathbb{E}[T_i^2] \lesssim \boldsymbol{y}_i^4/\rho^3 \le \mu^2 d^2/\rho^3 n^2$, and $|T_i| \lesssim \boldsymbol{y}_i^2/\rho^2 \le \mu d/\rho^2 n$. Subsequently, with probability $1 - O(N^{-10})$ we have

$$\left|\sum_{i=1}^{n} T_i\right| \lesssim \frac{\mu d}{\rho^{3/2}}\sqrt{\frac{\log^2 N}{n}}.$$

The estimation error of $(1-\rho)(\mathbf{X}^\top\mathbf{X})_{ii}$ for $(1-\rho)/\rho \cdot \|\boldsymbol{y}_i\|_2^2 = (1-\rho)/\rho$ can be upper bounded similarly. □

Proof of Lemma 9.4 Take $\boldsymbol{\Delta}_{jk} = 3\psi_1\sqrt{\frac{\log N}{n}}$ for $j \neq k$ and $\boldsymbol{\Delta}_{jk} = 3\psi_2\sqrt{\frac{\log N}{n}}$. Fix arbitrary $\tilde{\boldsymbol{x}}_j \notin \mathcal{S}_\ell$ and $\tilde{\boldsymbol{x}}_i \in \mathcal{S}_\ell$. Because $\tilde{\boldsymbol{\nu}}_i = \lambda(\tilde{\boldsymbol{x}}_i - \widetilde{\mathbf{X}}_{-i}^{(\ell)}\tilde{\boldsymbol{c}}_i)$, we have that

$$\begin{aligned}\left|\langle\tilde{\boldsymbol{x}}_j, \tilde{\boldsymbol{\nu}}_i\rangle\right| &= \lambda\left|\tilde{\boldsymbol{x}}_j^\top(\tilde{\boldsymbol{x}}_i + \widetilde{\mathbf{X}}_{-i}^{(\ell)}\tilde{\boldsymbol{c}}_i)\right| \le \lambda(1+\|\tilde{\boldsymbol{c}}_i\|_1)\cdot \sup_{\tilde{\boldsymbol{x}}_i\in\mathcal{S}_\ell}\left|\langle\tilde{\boldsymbol{x}}_j, \tilde{\boldsymbol{x}}_i\rangle\right| \\ &\le \lambda(1+\|\tilde{\boldsymbol{c}}_i\|_1)\cdot\left(\tilde{\mu} + \sup_{\tilde{\boldsymbol{x}}_i\notin\mathcal{S}_\ell}\left|\langle\tilde{\boldsymbol{x}}_j, \tilde{\boldsymbol{x}}_i\rangle - \langle\boldsymbol{y}_j, \boldsymbol{y}_i\rangle\right|\right) \\ &\lesssim \lambda(1+\|\tilde{\boldsymbol{c}}_i\|_1)\cdot\left(\tilde{\mu} + \psi_1\sqrt{\frac{\log N}{n}}\right),\end{aligned} \tag{9.8}$$

where the last inequality holds by applying Definition 9.5 and the fact that

$$\begin{aligned}\left|\langle\tilde{\boldsymbol{x}}_i, \tilde{\boldsymbol{x}}_j\rangle - \langle\tilde{\boldsymbol{y}}_i, \tilde{\boldsymbol{y}}_j\rangle\right| &\le \left|(\widetilde{\boldsymbol{\Sigma}}_+)_{ij} - (\widetilde{\boldsymbol{\Sigma}})_{ij}\right| + \left|(\widetilde{\boldsymbol{\Sigma}})_{ij} - \langle\tilde{\boldsymbol{y}}_i, \tilde{\boldsymbol{y}}_j\rangle\right| \\ &\le \left|\boldsymbol{\Delta}_{ij}\right| + \left|\langle\tilde{\boldsymbol{x}}_i, \tilde{\boldsymbol{x}}_j\rangle - \langle\tilde{\boldsymbol{y}}_i, \tilde{\boldsymbol{y}}_j\rangle\right| \\ &\le \left|\boldsymbol{\Delta}_{ij}\right| + \left|\langle\tilde{\boldsymbol{z}}_i, \tilde{\boldsymbol{y}}_j\rangle\right| + \left|\langle\tilde{\boldsymbol{y}}_j, \tilde{\boldsymbol{z}}_i\rangle\right| + \left|\langle\tilde{\boldsymbol{z}}_j, \tilde{\boldsymbol{z}}_i\rangle\right| \\ &\lesssim \psi_1\sqrt{\frac{\log N}{n}} \quad \text{for } i \neq j.\end{aligned}$$

To bound $\|\tilde{\boldsymbol{c}}_i\|_1$, consider an auxiliary noiseless problem:

$$\hat{\boldsymbol{c}}_i := \arg\min_{\boldsymbol{c}_i} \|\boldsymbol{c}_i\|_1 \quad s.t. \;\; \boldsymbol{y}_i = \mathbf{Y}_{-i}^{(\ell)}\boldsymbol{c}_i. \tag{9.9}$$

Note that when $r > 0$ Eq. (9.9) is always feasible. Following standard analysis (e.g., Lemma 15 and Eq. (5.15) of [WX16]), it can be established that $\|\hat{\boldsymbol{c}}_i\|_1 \le 1/r_i \le 1/r$. On the other hand, by optimality we have $\|\tilde{\boldsymbol{c}}_i\|_1 + \frac{\lambda}{2}\|\tilde{\boldsymbol{x}}_i - \widetilde{\mathbf{X}}_{-i}^{(\ell)}\tilde{\boldsymbol{c}}_i\|_2^2 \le \|\hat{\boldsymbol{c}}_i\|_1 + \frac{\lambda}{2}\|\tilde{\boldsymbol{x}}_i - \widetilde{\mathbf{X}}_{-i}^{(\ell)}\hat{\boldsymbol{c}}_i\|_2^2$. Therefore,

$$\begin{aligned}\|\tilde{\boldsymbol{c}}_i\|_1 &\le \|\hat{\boldsymbol{c}}_i\|_1 + \frac{\lambda}{2}\left\|\tilde{\boldsymbol{x}}_i - \widetilde{\mathbf{X}}_{-i}^{(\ell)}\hat{\boldsymbol{c}}_i\right\|_2^2 \\ &\lesssim \|\hat{\boldsymbol{c}}_i\|_1 + \frac{\lambda}{2}\left\|\boldsymbol{y}_i - \mathbf{Y}_{-i}^{(\ell)}\hat{\boldsymbol{c}}_i\right\|_2^2 + (1+\|\hat{\boldsymbol{c}}_i\|_1)^2\cdot\frac{\lambda}{2}\sup_{\boldsymbol{y}_i,\boldsymbol{y}_j\in\mathcal{S}_\ell}\left|\langle\tilde{\boldsymbol{x}}_i, \tilde{\boldsymbol{x}}_j\rangle - \langle\boldsymbol{y}_i, \boldsymbol{y}_j\rangle\right| \\ &= \|\hat{\boldsymbol{c}}_i\|_1 + (1+\|\hat{\boldsymbol{c}}_i\|_1)^2\cdot\frac{\lambda}{2}\sup_{\boldsymbol{y}_i,\boldsymbol{y}_j\in\mathcal{S}_\ell}\left|\langle\tilde{\boldsymbol{x}}_i, \tilde{\boldsymbol{x}}_j\rangle - \langle\boldsymbol{y}_i, \boldsymbol{y}_j\rangle\right|\end{aligned}$$

$$
\begin{aligned}
&\lesssim \|\hat{\boldsymbol{c}}_i\|_1 + (1+\|\hat{\boldsymbol{c}}_i\|_1)^2 \cdot (\psi_1+\psi_2)\sqrt{\frac{\log N}{n}} \\
&\lesssim \frac{1}{r}\left(1+\frac{\lambda}{r}(\psi_1+\psi_2)\sqrt{\frac{\log N}{n}}\right).
\end{aligned} \tag{9.10}
$$

□

Proof of Theorem 9.4 Following the analysis of Lasso SSC solution path in [WX16], it suffices to show that $\lambda > 1/\|\tilde{\boldsymbol{x}}_i^\top \widetilde{\mathbf{X}}_{-i}\|_\infty$. On the other hand, note that $\|\boldsymbol{y}_i^\top \mathbf{Y}_{-i}\|_\infty \geq \|\boldsymbol{y}_i^\top \mathbf{Y}_{-i}^{(\ell)}\|_\infty \geq r_i \geq r$ (see, for example, Eq. (5.19) of [WX16]). Subsequently,

$$
\left\|\tilde{\boldsymbol{x}}_i^\top \widetilde{\mathbf{X}}_{-i}\right\|_\infty \geq \left\|\boldsymbol{y}_i^\top \mathbf{Y}_{-i}\right\|_\infty - \sup_{j\neq i}\left|\langle\tilde{\boldsymbol{x}}_i,\tilde{\boldsymbol{x}}_j\rangle - \langle\boldsymbol{y}_i,\boldsymbol{y}_j\rangle\right| \gtrsim r - \psi_1\sqrt{\frac{\log N}{n}}.
$$

□

Proof of Lemma 9.5 We first prove

$$
\max_{\boldsymbol{y}_i\in\mathcal{S}_k}\max_{\boldsymbol{y}_j\in\mathcal{S}_\ell}\left|\langle\boldsymbol{y}_i,\boldsymbol{y}_j\rangle\right| \lesssim \chi_{k\ell}\cdot\frac{\log(\overline{C}N/\underline{C})}{\sqrt{d_k d_\ell}} \qquad \forall j\neq k\in[L]. \tag{9.11}
$$

Let N_k and N_ℓ be the total number of data points in $\mathcal{S}_k$ and $\mathcal{S}_\ell$, and let P_k and P_ℓ be the corresponding densities which are bounded from both above and below by $\overline{C}p_0$ and $\underline{C}p_0$. Consider a rejection sampling procedure: first sample $\boldsymbol{\alpha}$ randomly from the uniform measure over $\{\boldsymbol{\alpha}\in\mathbb{R}^{d_k}:\|\boldsymbol{\alpha}\|_2=1\}$, and then reject the sample if $u > p_k(\boldsymbol{\alpha})/\overline{C}p_0$, where $u\sim U(0,1)$. Repeat the procedure until N_k samples are obtained. This procedure is sound because $p_k/p_0\leq\overline{C}$, and the resulting (accepted) samples are i.i.d. distributed according to P_k. On the other hand, for any $\boldsymbol{\alpha}$ the probability of acceptance is lower bounded by $\underline{C}/\overline{C}$. Therefore, the procedure terminates by producing a total of $O(\overline{C}N_k/\underline{C})$ samples (both accepted and rejected). Thus, without loss of generality we can assume both P_k and P_ℓ are uniform measures on the corresponding spheres, by paying the cost of adding $\tilde{N}_k = O(\overline{C}N_k/\underline{C})$ and $\tilde{N}_\ell = O(\overline{C}N_\ell/\underline{C})$ points to each subspace.

Now fix $\boldsymbol{y}_i = \mathbf{U}_k\boldsymbol{\alpha}_i$ and $\boldsymbol{y}_j = \mathbf{U}_\ell\boldsymbol{\alpha}_j$, where $\boldsymbol{\alpha}_i\in\mathbb{R}^{d_k}$, $\boldsymbol{\alpha}_j\in\mathbb{R}^{d_\ell}$, and $\|\boldsymbol{\alpha}_i\|_2 = \|\boldsymbol{\alpha}_j\|_2 = 1$. Then both $\boldsymbol{\alpha}_i$ and $\boldsymbol{\alpha}_j$ are uniformly distributed on the low-dimensional spheres, and that $|\langle\boldsymbol{y}_i,\boldsymbol{y}_j\rangle| = |\boldsymbol{\alpha}_i^\top(\mathbf{U}_k^\top\mathbf{U}_\ell)\boldsymbol{\alpha}_j|$. Applying Lemma 7.5 of [SC12] and note that $\chi_{k\ell} = \|\mathbf{U}_k^\top\mathbf{U}_\ell\|_F$ we complete the proof of Eq. (9.11).

We next prove

$$
r_i \gtrsim \sqrt{\frac{\log(\underline{C}N_\ell/\overline{C}d_\ell)}{d_\ell}} \qquad \forall i\in[N], \ell\in[L], \boldsymbol{x}_i\in\mathcal{S}_\ell. \tag{9.12}
$$

Let P_ℓ be the underlying measure of subspace $\mathcal{S}_\ell$. Consider the decomposition $P_\ell = \underline{C}/\overline{C} \cdot P_0 + (1 - \underline{C}/\overline{C}) \cdot P'_\ell$, where P_0 is the uniform measure. Such a decomposition and the corresponding density P'_ℓ exist because $\underline{C} P_0 \le P_\ell \le \overline{C} P_0$. This shows that the distribution of points in subspace $\mathcal{S}_\ell$ can be expressed as a mixture distribution, with a uniform density mixture with weight probability $\underline{C}/\overline{C}$. Because r_i decreases with smaller data set, it suffices to consider only the uniform mixture. Thus, we can assume P_ℓ is the uniform measure at the cost of considering only $\tilde{N}_\ell = \Omega(\underline{C} N_\ell/\overline{C})$ points in subspace $\mathcal{S}_\ell$. Applying Lemma 21 of [WX16] and replacing N_ℓ with $\tilde{N}_\ell$ we complete the proof of Eq. (9.12).

Finally Lemma 9.5 is an easy corollary of Eqs. (9.11) and (9.12). □

Proof of Lemma 9.6 Fix $k, \ell \in [L]$ and let $\mathbf{U}_k = (\boldsymbol{u}_{k1}, \cdots, \boldsymbol{u}_{kd})$, $\mathbf{U}_\ell = (\boldsymbol{u}_{\ell 1}, \cdots, \boldsymbol{u}_{\ell d})$ be orthonormal basis of $\mathcal{U}_k$ and $\mathcal{U}_\ell$. Then $\chi_{k\ell} = \|\mathbf{U}_k^\top \mathbf{U}_\ell\|_F \le d\|\mathbf{U}_k^\top \mathbf{U}_\ell\|_{\max} = d \cdot \sup_{1\le i,j\le d} |\langle \boldsymbol{u}_{ki}, \boldsymbol{u}_{\ell j}\rangle|$. Because $\mathcal{U}_k$ and $\mathcal{U}_\ell$ are random subspaces, $\boldsymbol{u}_{ki}$ and $\boldsymbol{u}_{\ell j}$ are independent vectors distributed uniformly on the d-dimensional unit sphere. Applying Lemma 17 of [WX16] and a union bound over all i, j, k, ℓ we prove the upper bound on χ. For the upper bound on μ, simply note that $\|\boldsymbol{u}_{jk}\|_\infty \lesssim \sqrt{\frac{\log N}{n}}$ with probability $1 - O(N^{-10})$ by standard concentration result for Gaussian suprema. □

9.5 Concluding Remarks

Our numerical simulations first demonstrated the spectral clustering accuracy with respect to the effect of Gaussian noise. In this experiment, ambient dimension $n = 100$, intrinsic dimension $d = 4$, the number of clusters $L = 10$, the number of data points $N = 1000$, and the Gaussian noise is $Z_i j\ N(0, \sigma/\sqrt{n}$, where σ is changed from 0.00 to 1.00 with step length 0.01.

The second experiments investigated the RelViolation with respect to Gaussian noise σ and missing rate ρ. We change σ from 0 to 1 with step length 0.01 and set ρ as 0.03, 0.05, and 0.10, respectively. In these experiments, ambient dimension $n = 10$, intrinsic dimension $d = 2$, the number of clusters $L = 5$, and the number of data points $N = 100$.

Our last numerical simulations test the effects of Gaussian noise σ, subspace rand d, and number of clusters L, respectively.

An interesting future direction is to further improve the sample complexity to $\rho = \Omega(n^{-1/2})$ without knowing the norms $\|\boldsymbol{y}_i\|_2$. Such sample complexity is likely to be optimal because it is the smallest observation rate under which off-diagonal elements of sample covariance $\mathbf{X}^\top \mathbf{X}$ can be consistently estimated in max norm, which is also shown to be optimal for related regression problems [WWBS17].

Chapter 10
Online Discovery for Stable and Grouping Causalities in Multivariate Time Series

The content of this chapter is organized as follows: The problem formulation is presented in Sect. 10.1. Section 10.2 introduces the details about our proposed approach and its equivalent Bayesian model. A solution capable of online inference with particle learning is given in Sect. 10.3. Extensive empirical evaluation is demonstrated in Sect. 10.4. Finally, we conclude our work and discuss the future work.

10.1 Problem Formulation

In this section, we formally define the Granger causality by the VAR model. Given a set of time series $\mathbf{Y}$ defined on $\mathbb{R}^n$ in the time interval $[0, T]$, that is,

$$\mathbf{Y} = \{\mathbf{y}_t : \mathbf{y}_t \in \mathbb{R}^n, \ t \in [0, T]\},$$

where $\mathbf{y}_t = (y_{t,1}, y_{t,2}, \ldots, y_{t,n})^T$. The inference of Granger causality is usually achieved by fitting the time-series data $\mathbf{Y}$ with a VAR model. Given the maximum time lag L, the VAR model is expressed as follows:

$$\mathbf{y}_t = \sum_{l=1}^{L} W_l^T \mathbf{y}_{t-l} + \boldsymbol{\epsilon}, \tag{10.1}$$

Part of this chapter is in the paper titled "Online Discovery for Stable and Grouping Causalities in Multivariate Time Series" by Wentao Wang, Bin Shi et al. (2018) presently under review for publication.

B. Shi, S. S. Iyengar, *Mathematical Theories of Machine Learning - Theory and Applications*, https://doi.org/10.1007/978-3-030-17076-9_10

where ϵ is the standard Gaussian noise and the vector-value $\mathbf{y}_t$ $(1 \leq t \leq T)$ only depends on the past vector-value $\mathbf{y}_{t-l}$ $(1 \leq l \leq L)$. The Granger causal relationship between $\mathbf{y}_t$ and $\mathbf{y}_{t-l}$ is formulated as the matrix W_l in the following:

$$W_l = (w_{l,ji})_{n\times n},$$

where the entry $w_{l,ji}$ expresses how large the component $y_{t-l,i}$ influences the component $y_{t,j}$, noted as $y_{t-l,i} \rightarrow_g y_{t,j}$.

To induce sparsity in the matrix W_l $(l = 1, 2, \ldots, L)$, the prior work [ZWW+16] proposed a VAR-Lasso model as follows:

$$\min_{W_l} \sum_{t=L+1}^{T} \left\| \mathbf{y}_t - \sum_{l=1}^{L} W_l^T \mathbf{y}_{t-l} \right\|_2^2 + \lambda_1 \sum_{l=1}^{L} \|W_l\|_1, \tag{10.2}$$

and an online time-varying method based on Bayesian update. However, it suffers instability and fails to select a group of variables that are highly correlated. To address these problems, we propose a method with elastic-net regularization and an equivalent online inference strategy is given in the following sections.

10.2 Elastic-Net Regularizer

In this section, we describe the VAR-elastic-net model and its equivalent form in the perspective of Bayesian modeling.

10.2.1 Basic Optimization Model

Elastic-net regularization [ZH05] is a combination of L_1 and L_2 norm and has the following objective function for MTS data:

$$\sum_{t=L+1}^{T} \left\| \mathbf{y}_t - \sum_{l=1}^{L} W_l^T \mathbf{y}_{t-l} \right\|_2^2 + \lambda_1 \sum_{l=1}^{L} \|W_l\|_1 + \lambda_2 \sum_{l=1}^{L} \|W_l\|_2^2, \tag{10.3}$$

where $\|\cdot\|_1$ is the entrywise norm and $\|\cdot\|_2$ is the Frobenius norm (or Hilbert–Schmidt norm).

In order to change Eq. (10.3) into the standard form of the linear regression model, we define $\boldsymbol{\beta}$, a $nL \times n$ matrix, as follows:

$$\boldsymbol{\beta} = (W_1^T, W_2^T, \ldots, W_L^T)^T, \tag{10.4}$$

and $\mathbf{x}_t$ be a nL column vector as

$$\mathbf{x}_t = [\mathbf{y}_{t-1}^T, \mathbf{y}_{t-2}^T, \ldots, \mathbf{y}_{t-L}^T]^T. \tag{10.5}$$

Then, Eq. (10.3) can be reformulated as

$$\sum_{t=L+1}^{T} \left(\mathbf{y}_t - \boldsymbol{\beta}^T \mathbf{x}_t\right)^2 + \lambda_1 \|\boldsymbol{\beta}\|_1 + \lambda_2 \|\boldsymbol{\beta}\|_2^2. \tag{10.6}$$

The coefficient matrix $\boldsymbol{\beta}$ can be expressed as

$$\boldsymbol{\beta} = (\boldsymbol{\beta}_1^T, \boldsymbol{\beta}_2^T, \ldots, \boldsymbol{\beta}_n^T)^T, \tag{10.7}$$

where $\boldsymbol{\beta}_i$ $(i = 1, 2, \ldots, n)$ is a row vector of size nL. Based on Eq. (10.7), the equivalent form of Eq. (10.6) is

$$\sum_{t=L+1}^{T} \left(y_{t,i} - \boldsymbol{\beta}_i^T \mathbf{x}_t\right)^2 + \lambda_1 \|\boldsymbol{\beta}_i\|_1 + \lambda_2 \|\boldsymbol{\beta}_i\|_2^2, \tag{10.8}$$

where $i = 1, 2, \ldots, n$. Thus, the original optimization problem of Eq. (10.3) can be addressed as the optimization problem of n independent standard linear regression problem of Eq. (10.8).

10.2.2 The Corresponding Bayesian Model

From Bayesian perspective, $y_{t,i}$ $(i = 1, 2, \ldots, n)$ follows a Gaussian distribution, given the coefficient vector $\boldsymbol{\beta}_i$ and the variance of random observation noise σ_i^2, as follows:

$$y_{t,i} | \boldsymbol{\beta}_i, \sigma_i^2 \sim \mathcal{N}\left(\boldsymbol{\beta}_i^T \mathbf{x}_t, \sigma_i^2\right). \tag{10.9}$$

The coefficient vector $\boldsymbol{\beta}_i$ is viewed as a random variable which follows the mixed Gaussian and Laplace distribution, as below [LL+10, Mur12]:

$$p(\boldsymbol{\beta}_i | \sigma_i^2) \propto \exp\left(-\lambda_1 \sigma_i^{-1} \sum_{j=1}^{nL} |\boldsymbol{\beta}_{ij}| - \lambda_2 \sigma_i^{-2} \sum_{j=1}^{nL} |\boldsymbol{\beta}_{ij}^2|\right). \tag{10.10}$$

Equation (10.10) represents a scale mixture of normal distributions and exponential distributions and equals a hierarchical form as below:

$$\begin{aligned}
&\tau_j^2|\lambda_1 \sim \sqrt{\exp\left(\lambda_1^2\right)}, \\
&\boldsymbol{\beta}_i|\sigma_i^2, \tau_1^2, \ldots, \tau_{nL}^2 \sim \mathcal{N}\left(0, \sigma_i^2\mathbf{M}_{\boldsymbol{\beta}_i}\right), \\
&\mathbf{M}_{\boldsymbol{\beta}_i} = diag\left(\left(\lambda_2 + \tau_1^{-2}\right)^{-1}, \ldots, \left(\lambda_2 + \tau_{nL}^{-2}\right)^{-1}\right).
\end{aligned} \tag{10.11}$$

The variance σ_i^2 is a random variable following inverse gamma distribution [Mur12] as follows:

$$\sigma_i^2 \sim \mathcal{IG}(\alpha_1, \alpha_2), \tag{10.12}$$

where α_1 and α_2 are hyperparameters.

Equation (10.10) can be obtained from integrating out the hyperparameters α_1 and α_2 in Eq. (10.12) and it reduces to the regular Lasso when $\lambda_2 = 0$.

10.2.3 Time-Varying Causal Relationship Model

The aforementioned model is the traditional static regression model, based on the assumption that the coefficient $\boldsymbol{\beta}_i (i = 1, 2, \ldots, n)$ is unknown but fixed, which rarely holds in practice. To model dynamic dependencies, it is reasonable to view the coefficient vector $\boldsymbol{\beta}_{t,i} (i = 1, 2, \ldots, n)$ as a function of time t. Specifically, we propose a method for modeling the coefficient vector as two parts including the stationary part and the drift part. The latter is to account for tracking the time-varying temporal dependency among the time series instantly.

Let the operation $\circ$ be the Hadamard product (entrywise-product). The form of the dynamic coefficient vector $\boldsymbol{\beta}_{t,i}$ $(i = 1, 2, \ldots, n)$ is constructed as

$$\boldsymbol{\beta}_{t,i} = \boldsymbol{\beta}_{t,i,1} + \boldsymbol{\beta}_{t,i,2} \circ \boldsymbol{\eta}_{t,i}, \tag{10.13}$$

where both $\boldsymbol{\beta}_{t,i,1}$ and $\boldsymbol{\beta}_{t,i,2}$ are stationary part and $\boldsymbol{\eta}_{t,i}$ is the drift part. The drift part at time t is caused by the standard Gaussian random walk from the information at time $t-1$, i.e., $\boldsymbol{\eta}_{t,i} = \boldsymbol{\eta}_{t-1,i} + \mathbf{v}$ and $\mathbf{v} \sim \mathcal{N}(0, \mathbf{I}_{nL})$. Thus $\boldsymbol{\eta}_{t,i}$ follows the Gaussian distribution

$$\boldsymbol{\eta}_{t,i} \sim \mathcal{N}(\boldsymbol{\eta}_{t-1,i}, \mathbf{I}_{nL}). \tag{10.14}$$

Combined with Eq. (10.13), the equivalent time-varying Bayesian elastic-net model in Eq. (10.8) becomes

$$\sum_{t=L+1}^{T} \left(y_{t,i} - \boldsymbol{\beta}_{t,i}^T \mathbf{x}_t\right)^2 + \lambda_{1,1}\|\boldsymbol{\beta}_{t,i,1}\|_1 \\ + \lambda_{2,1}\|\boldsymbol{\beta}_{t,i,1}\|_2^2 + \lambda_{1,2}\|\boldsymbol{\beta}_{t,i,2}\|_1 + \lambda_{2,2}\|\boldsymbol{\beta}_{t,i,2}\|_2^2. \tag{10.15}$$

Furthermore, the priors of the equivalent Bayesian model are given as below:

$$\begin{aligned}
&\boldsymbol{\beta}_{i,1}|\sigma_i^2, \tau_{1,1}^2, \ldots, \tau_{1,nL}^2 \sim \mathcal{N}\left(0, \sigma_i^2 \mathbf{M}_{\boldsymbol{\beta}_{i,1}}\right), \\
&\boldsymbol{\beta}_{i,2}|\sigma_i^2, \tau_{2,1}^2, \ldots, \tau_{2,nL}^2 \sim \mathcal{N}\left(0, \sigma_i^2 \mathbf{M}_{\boldsymbol{\beta}_{i,2}}\right), \\
&\tau_{1,j}^2|\lambda_{1,1} \sim \sqrt{\exp\left(\lambda_{1,1}^2\right)}, \\
&\tau_{2,j}^2|\lambda_{1,2} \sim \sqrt{\exp\left(\lambda_{1,2}^2\right)}, \\
&\sigma_i^2 \sim \mathcal{IG}(\alpha_1, \alpha_2), \\
&\mathbf{M}_{\boldsymbol{\beta}_{i,1}} = diag\left((\lambda_{2,1} + \tau_{1,1}^{-2})^{-1}, \ldots, (\lambda_{2,1} + \tau_{1,nL}^{-2})^{-1}\right), \\
&\mathbf{M}_{\boldsymbol{\beta}_{i,2}} = diag\left((\lambda_{2,2} + \tau_{2,1}^{-2})^{-1}, \ldots, (\lambda_{2,2} + \tau_{2,nL}^{-2})^{-1}\right).
\end{aligned} \tag{10.16}$$

It is difficult to solve straightforward the above regression model by traditional optimization method. We develop our solution to infer VAR-elastic-net model from a Bayesian perspective utilizing particle learning, which is presented in the following section.

10.3 Online Inference

Inference in general is the act or process of deriving logical conclusions from known premises or values that are assumed to be true. Our goal in this section of the book is to use the technique of online inference to infer both the latent parameters and the state variables in our Bayesian model. However, since the inference partially depends on the random walk which generates the latent state variables, we use particle learning strategy [CJLP10] to learn the distribution of both parameters and state variables.

Thus, here we describe the online inference process to be used to update the parameters from time $t-1$ to time t based on particle learning. At last, we give the pseudocode of algorithm to summarize the whole process.

The definition of a particle is as given below.

Definition 10.1 (Particle) A particle used to predict $y_{t,i}$ $(i = 1, 2, \ldots, n)$ is a container which maintains the current status information for value prediction. The status information comprises of random variables and their distributions with the corresponding hyperparameters.

Assume the number of particles is B. Let $\mathcal{P}_{t,i}^{(k)}$ be the kth particle for predicting the value y_i at time t with particle weight $\rho_{t,i}^{(k)}$.

We define a new variable $\boldsymbol{\beta}_{t,i}^{\prime(k)} = \left(\boldsymbol{\beta}_{t,i,1}^{(k),T}, \boldsymbol{\beta}_{t,i,2}^{(k),T}\right)^T$ for concisely expressing the stationary parts given in Eq. (10.13). At time $t-1$, the information of particle $\mathcal{P}_{t-1,i}^{(k)}$ includes the following variables and hyperparameters for the corresponding distributions:

$$
\begin{aligned}
\boldsymbol{\beta}_{t-1,i}^{\prime(k)} &\sim \mathcal{N}\left(\boldsymbol{\mu}_{\boldsymbol{\beta}_{t-1,i}^{\prime(k)}}, \sigma_i^2 \mathbf{M}^{\frac{1}{2}}_{\boldsymbol{\beta}_{t-1,i}^{\prime(k)}} \boldsymbol{\Sigma}_{\boldsymbol{\beta}_{t-1,i}^{\prime(k)}} \mathbf{M}^{\frac{1}{2}}_{\boldsymbol{\beta}_{t-1,i}^{\prime(k)}}\right), \\
\boldsymbol{\eta}_{t-1,i}^{(k)} &\sim \mathcal{N}\left(\boldsymbol{\mu}_{\boldsymbol{\eta}_{t-1,i}^{(k)}}, \boldsymbol{\Sigma}_{\boldsymbol{\eta}_{t-1,i}^{(k)}}\right), \\
\sigma^{2(k)}_{t-1,i} &\sim \mathcal{IG}\left(\alpha_{t-1,1}^{(k)}, \alpha_{t-1,2}^{(k)}\right).
\end{aligned}
\tag{10.17}
$$

10.3.1 Particle Learning

Particle learning as described by previous works in this field [CJLP10] is seen as a very powerful tool that can be used to provide an online inference strategy when working with Bayesian models. It falls under the broad category of sequential Monte Carlo (SMC) methods which in turn within it comprises a set of Monte Carlo methodologies that can be used in solving the filtering problem. It can be noted that particle learning also provides state filtering, sequential parameter learning, and smoothing in a general class of state space models.

The core idea behind the use of particle learning is the creation of a particle algorithm that can be directly applied on samples from the particle approximation to the joint posterior distribution of states and conditional sufficient statistics for fixed parameters in a fully adapted resample-propagate framework. This idea for particle learning is iterated in the following steps:

(1) At time $t-1$, there are B particles, and each contains information in Eq. (10.17). The coefficients at $t-1$ is given as

$$
\boldsymbol{\beta}_{t-1,i}^{(k)} = \boldsymbol{\beta}_{t-1,i,1}^{(k)} + \boldsymbol{\beta}_{t-1,i,2}^{(k)} \circ \boldsymbol{\eta}_{t-1,i}^{(k)}.
$$

(2) At time t, sample the drift part $\boldsymbol{\eta}_{t,i}^{(k)}$ from Eq. (10.14), and update parameters of all priors and sample the new values for $\boldsymbol{\beta}_{t,i,1}^{(k)}$, $\boldsymbol{\beta}_{t,i,2}^{(k)}$ (details are given in Sect. 10.3.2) for each particle.
(3) Finally, gain new feedback $y_{t,i}$ and resample B particles based on the recalculated particle weights (details are given in Sect. 10.3.2). The value of $\boldsymbol{\beta}_{t,i}$ for prediction at time t is averaged as below:

$$\boldsymbol{\beta}_{t,i} = \frac{1}{B}\sum_{k=1}^{B}\left(\boldsymbol{\beta}_{t,i,1}^{(k)} + \boldsymbol{\beta}_{t,i,2}^{(k)} \circ \boldsymbol{\eta}_{t,i}^{(k)}\right). \tag{10.18}$$

10.3.2 *Update Process*

In the process of particle learning, the key step is to update all the parameters from time $t-1$ to time t and recalculate particle weights mentioned above. In this section, we describe the update process of particle weights and all the parameters in detail.

Particle Weights Update
Each particle $\mathcal{P}_{t,i}^{(k)}$ has a weight, denoted as $\rho_{t,i}^{(k)}$, indicating its fitness for the new observed data at time t. Note that $\sum_{k=1}^{B}\rho_{t,i}^{(k)} = 1$. The fitness of each particle $\mathcal{P}_{t,i}^{(k)}$ is defined as likelihood of the observed data $\mathbf{x}_t$ and $y_{t,i}$. Therefore,

$$\rho_{t,i}^{(k)} \propto P(\mathbf{x}_t, y_{t,i}|\mathcal{P}_{t-1,i}^{(k)}).$$

Combined with Eq. (10.14) for $\boldsymbol{\eta}_{t,i}^{(k)}$ and Eq. (10.17) for $\boldsymbol{\eta}_{t-1,i}^{(k)}$, the particle weights $\rho_i^{(k)}$ at time t is proportional to the value as follows:

$$\begin{aligned}\rho_{t,i}^{(k)} \propto \iint & \mathcal{N}(y_{t,i}|\boldsymbol{\beta}_{t,i}^{(k),T}\mathbf{x}_t, \sigma_{t-1,i}^{2(k)})\mathcal{N}(\boldsymbol{\eta}_{t,i}^{(k)}|\boldsymbol{\eta}_{t-1,i}^{(k)}, I_{nL}) \\ & \mathcal{N}(\boldsymbol{\eta}_{t-1,i}^{(k)}|\boldsymbol{\mu}_{\boldsymbol{\eta}_{t-1,i}^{(k)}}, \boldsymbol{\Sigma}_{\boldsymbol{\eta}_{t-1,i}^{(k)}})d\boldsymbol{\eta}_{t-1,i}^{(k)}d\boldsymbol{\eta}_{t,i}^{(k)}.\end{aligned} \tag{10.19}$$

Integrating out the variables of $\boldsymbol{\eta}_{t-1,i}^{(k)}$ and $\boldsymbol{\eta}_{t,i}^{(k)}$, we can obtain that the particle weights $\rho_i^{(k)}$ at time t follow Gaussian distribution as below:

$$\rho_{t,i}^{(k)} \propto \mathcal{N}(y_{t,i}|m_{t,i}^{(k)}, Q_{t,i}^{(k)}), \tag{10.20}$$

where the mean value and the variance are, respectively,

$$
\begin{aligned}
m_{t,i}^{(k)} &= (\boldsymbol{\beta}_{t,i,1}^{(k)} + \boldsymbol{\beta}_{t,i,2}^{(k)} \circ \boldsymbol{\mu}_{\boldsymbol{\eta}_{t-1,i}^{(k)}})^T \mathbf{x}_t, \\
Q_{t,i}^{(k)} &= \sigma_{t-1,i}^{2(k)} + (\mathbf{x}_t \circ \boldsymbol{\beta}_{t,i,2}^{(k)})^T \\
&\qquad (\mathbf{I}_{nL} + \boldsymbol{\Sigma}_{\boldsymbol{\eta}_{t-1,i}^{(k)}})(\mathbf{x}_t \circ \boldsymbol{\beta}_{t,i,2}^{(k)}).
\end{aligned} \tag{10.21}
$$

Furthermore, the final k^{th} particle weights at time t can be obtained from normalization, as follows:

$$
\rho_{t,i}^{(k)} = \frac{\mathcal{N}(y_{t,i}|m_{t,i}^{(k)}, Q_{t,i}^{(k)})}{\sum_{k=1}^{B} \mathcal{N}(y_{t,i}|m_{t,i}^{(k)}, Q_{t,i}^{(k)})}. \tag{10.22}
$$

With the particle weights $\rho_{t,i}^{(k)}$ $(k = 1, 2, \ldots, B)$ at time t obtained, the B particles are resampled at time t.

Latent State Update

Having the new observation $\mathbf{x}_t$ and $y_{t,i}$ at time t, both the mean $\boldsymbol{\mu}_{\boldsymbol{\eta}_{t-1,i}^{(k)}}$ and the variance $\boldsymbol{\Sigma}_{\boldsymbol{\eta}_{t-1,i}^{(k)}}$ need to update from time $t-1$ to time t. Here, we apply the Kalman filter method [Har90] to recursively update to the mean and the variance at time t as follows:

$$
\begin{aligned}
\boldsymbol{\mu}_{\boldsymbol{\eta}_{t,i}^{(k)}} &= \boldsymbol{\mu}_{\boldsymbol{\eta}_{t-1,i}^{(k)}} + \mathbf{G}_{t-1,i}^{(k)} \\
&\quad \left(y_{t,i} - \left(\boldsymbol{\beta}_{t,i,1}^{(k)} + \boldsymbol{\beta}_{t,i,2}^{(k)} \circ \boldsymbol{\mu}_{\boldsymbol{\eta}_{t-1,i}^{(k)}} \right)^T \mathbf{x}_t \right), \\
\boldsymbol{\Sigma}_{\boldsymbol{\eta}_{t,i}^{(k)}} &= \boldsymbol{\Sigma}_{\boldsymbol{\eta}_{t-1,i}^{(k)}} + I_{nL} - \mathbf{G}_{t,i}^{(k)} Q_{t,i}^{(k)} \mathbf{G}_{t,i}^{(k),T}.
\end{aligned} \tag{10.23}
$$

where $\mathbf{G}_{t,i}^{(k)}$ is the Kalman gain defined as [Har90]

$$
\mathbf{G}_{t,i}^{(k)} = \left(\mathbf{I}_{nL} + \boldsymbol{\Sigma}_{\boldsymbol{\eta}_{t-1,i}^{(k)}} \right) \left(\mathbf{x}_t \circ \boldsymbol{\beta}_{t-1,i,2}^{\prime(k)} \right) Q_{t,i}^{(k)^{-1}}. \tag{10.24}
$$

Then, we can sample the drift part at time t from Gaussian distribution as follows:

$$
\boldsymbol{\eta}_{t,i}^{(k)} \sim \mathcal{N}\left(\boldsymbol{\mu}_{\boldsymbol{\eta}_{t,i}^{(k)}}, \boldsymbol{\Sigma}_{\boldsymbol{\eta}_{t,i}^{(k)}} \right). \tag{10.25}
$$

Before updating parameters, a resampling process is conducted. We replace the particle set $\mathcal{P}_{t-1,i}^{(k)}$ with a new set $\mathcal{P}_{t,i}^{(k)}$, where $\mathcal{P}_{t,i}^{(k)}$ is generated from $\mathcal{P}_{t-1,i}^{(k)}$ using sampling with replacement based on the new particle weights.

Parameter Update

Having sampled the drift part $\boldsymbol{\eta}_{t,i}^{(k)}$, the parameters update for the covariant matrix, mean value, and the hyperparameters from time $t-1$ to time t is as follows:

$$\begin{aligned}
\boldsymbol{\Sigma}_{\boldsymbol{\beta}_{t,i}^{\prime(k)}} &= \left(\boldsymbol{\Sigma}_{\boldsymbol{\beta}_{t-1,i}^{\prime(k)}}^{-1} + \mathbf{M}_{\boldsymbol{\beta}_{t-1,i}^{\prime(k)}}^{\frac{1}{2}} \mathbf{z}_{t,i}^{(k)} \mathbf{z}_{t,i}^{(k)T} \mathbf{M}_{\boldsymbol{\beta}_{t-1,i}^{\prime(k)}}^{\frac{1}{2}}\right)^{-1}, \\
\boldsymbol{\mu}_{\boldsymbol{\beta}_{t,i}^{\prime(k)}} &= \mathbf{M}_{\boldsymbol{\beta}_{t-1,i}^{\prime(k)}}^{\frac{1}{2}} \boldsymbol{\Sigma}_{\boldsymbol{\beta}_{t,i}^{\prime(k)}} \mathbf{M}_{\boldsymbol{\beta}_{t-1,i}^{\prime(k)}}^{\frac{1}{2}} \mathbf{z}_{t,i}^{(k)} y_{t,i} \\
&\quad + \mathbf{M}_{\boldsymbol{\beta}_{t-1,i}^{\prime(k)}}^{\frac{1}{2}} \boldsymbol{\Sigma}_{\boldsymbol{\beta}_{t,i}^{\prime(k)}} \boldsymbol{\Sigma}_{\boldsymbol{\beta}_{t-1,i}^{\prime(k)}} \mathbf{M}_{\boldsymbol{\beta}_{t-1,i}^{\prime(k)}}^{\frac{1}{2}} \boldsymbol{\beta}_{t-1,i}^{\prime(k)}, \\
\alpha_{t,1}^{(k)} &= \alpha_{t-1,1}^{(k)} + \frac{1}{2}, \\
\alpha_{t,2}^{(k)} &= \alpha_{t-1,2}^{(k)} + \frac{1}{2} y_{t,i}^2 \\
&\quad + \frac{1}{2} \boldsymbol{\mu}_{\boldsymbol{\beta}_{t-1,i}^{\prime(k)}}^T \mathbf{M}_{\boldsymbol{\beta}_{t-1,i}^{\prime(k)}}^{-\frac{1}{2}} \boldsymbol{\Sigma}_{\boldsymbol{\beta}_{t-1,i}^{\prime(k)}} \mathbf{M}_{\boldsymbol{\beta}_{t-1,i}^{\prime(k)}}^{-\frac{1}{2}} \boldsymbol{\mu}_{\boldsymbol{\beta}_{t-1,i}^{\prime(k)}} \\
&\quad - \frac{1}{2} \boldsymbol{\mu}_{\boldsymbol{\beta}_{t,i}^{\prime(k)}}^T \mathbf{M}_{\boldsymbol{\beta}_{t,i}^{\prime(k)}}^{-\frac{1}{2}} \boldsymbol{\Sigma}_{\boldsymbol{\beta}_{t,i}^{\prime(k)}} \mathbf{M}_{\boldsymbol{\beta}_{t,i}^{\prime(k)}}^{-\frac{1}{2}} \boldsymbol{\mu}_{\boldsymbol{\beta}_{t,i}^{\prime(k)}},
\end{aligned} \tag{10.26}$$

where $\mathbf{z}_{t,i}^{(k)} = (\mathbf{x}_t^T, (\boldsymbol{\eta}_{t,i}^{(k)} \circ \mathbf{x}_t)^T)^T$ be a $2n$ column vector. After parameters update in Eq. (10.26) at time t, we can sample $\sigma^{2(k)}_{t,i}$ and the stationary part of coefficients $\boldsymbol{\beta}_{t,i}^{\prime(k)}$ as follows:

$$\begin{aligned}
\sigma^{2(k)}_{t,i} &\sim \mathcal{IG}(\alpha_{t,1}^{(k)}, \alpha_{t,2}^{(k)}), \\
\boldsymbol{\beta}_{t,i}^{\prime(k)} &\sim \mathcal{N}\left(\boldsymbol{\mu}_{\boldsymbol{\beta}_{t,i}^{\prime(k)}}, \sigma^{2(k)}_{t,i} \mathbf{M}_{\boldsymbol{\beta}_{t,i}^{\prime(k)}}^{\frac{1}{2}} \boldsymbol{\Sigma}_{\boldsymbol{\beta}_{t,i}^{\prime(k)}} \mathbf{M}_{\boldsymbol{\beta}_{t,i}^{\prime(k)}}^{\frac{1}{2}}\right).
\end{aligned} \tag{10.27}$$

10.3.3 Algorithm

Putting all the aforementioned descriptions together, an algorithm for VAR-elastic-net by Bayesian update is provided below.

Online inference for time-varying Bayesian VAR-elastic-net model starts with MAIN procedure, as presented in Algorithm 1. The parameters B, L, α_1, α_2, λ_{11}, λ_{12}, λ_{21}, and λ_{22} are given as the input of MAIN procedure. The initialization is executed from line 2 to line 7. As new observation $\mathbf{y}_t$ arrives at time t, $\mathbf{x}_t$ is built using the time lag, then $\boldsymbol{\beta}_t$ is inferred by calling UPDATE procedure. Especially in the UPDATE procedure, we use the resample-propagate strategy in particle learning [CJLP10] rather than the resample-propagate strategy in particle filtering [DKZ+03]. With the resample-propagate strategy, the particles are resampled by taking $\rho_{t,i}^{(k)}$ as the kth particle's weight, where the $\rho_{t,i}^{(k)}$ indicates the occurring probability of the observation at time t given the particle at time $t-1$. The resample-propagate strategy is considered as an optimal and fully adapted strategy, avoiding an important sampling step.

Algorithm 1 The algorithm for VAR-elastic-net by Bayesian update

1: **procedure** MAIN($B, L, \alpha_1, \alpha_2, \lambda_{11}, \lambda_{12}, \lambda_{21}, \lambda_{22}, \boldsymbol{\beta}_t$)
2:     **for** $i = 1 : K$ **do**
3:         Initialize $y_{0,i}$ with B particles;
4:         **for** $k = 1 : B$ **do**
5:             Initialize $\boldsymbol{\mu}_{\boldsymbol{\beta}_{0,i}^{\prime(k)}} = \mathbf{0}$;
6:             Initialize $\boldsymbol{\Sigma}_{\boldsymbol{\beta}_{0,i}^{\prime(k)}} = \mathbf{I}$;
7:         **end for**
8:     **end for**
9:     **for** $t = 1 : T$ **do**
10:         Obtain $\mathbf{x}_t$ using time lag L;
11:         **for** $i = 1 : K$ **do**
12:             UPDATE($\mathbf{x}_t, y_{t,i}, \boldsymbol{\beta}_{t,i}^{\prime}, \boldsymbol{\eta}_{t,i}$);
13:             Output $\boldsymbol{\beta}_t$ according to Eq. (10.18);
14:         **end for**
15:     **end for**
16: **end procedure**

17: **procedure** UPDATE($\mathbf{x}_t, y_{t,i}, \boldsymbol{\beta}_{t,i}^{\prime}, \boldsymbol{\eta}_{t,i}$)
18:     **for** $k = 1 : B$ **do**
19:         Compute particle weights $\rho_{t,i}^{(k)}$ by Eq. (10.22);
20:     **end for**
21:     Resample $\mathcal{P}_{t,i}^{(k)}$ from $\mathcal{P}_{t-1,i}^{(k)}$ according to $\rho_{t,i}^{(k)}$;
22:     **for** $i = 1 : B$ **do**
23:         Update $\boldsymbol{\mu}_{\boldsymbol{\eta}_{t,i}^{(k)}}$ and $\boldsymbol{\Sigma}_{\boldsymbol{\eta}_{t,i}^{(k)}}$ by Eq. (10.23);
24:         Sample $\boldsymbol{\eta}_{t,i}^{(k)}$ according to Eq. (10.25);
25:         Update the parameters $\boldsymbol{\beta}_{t,i}^{\prime(k)}, \boldsymbol{\beta}_{t,i}^{\prime(k)}, \alpha_{t,1}^{(k)}$ and $\alpha_{t,2}^{(k)}$ by Eq. (10.26);
26:         Sample $\sigma^{2}{}_{t,i}^{(k)}$ and $\boldsymbol{\beta}_{t,i}^{\prime(k)}$ by Eq. (10.27);
27:     **end for**
28: **end procedure**

10.4 Empirical Study

To demonstrate the efficiency of our proposed algorithm, we conduct experiments over both synthetic and real-world climate change data set. In this section, we first outline the baseline algorithms for comparison and the evaluation metrics. Second, we present our approach to generate the synthetic data and then illustrate the corresponding experimental results in detail. Finally, a case study on real-world climate change data set is given.

10.4.1 Baseline Algorithms

In our experiments, we demonstrate the performance of our method by comparing with the following baseline algorithms:

- $BL(\gamma)$: VAR by Bayesian prior Gaussian distribution $\mathcal{N}(\mathbf{0}, \gamma^{-1}\mathbf{I}_d)$.
- $BLasso(\lambda_1)$: VAR-Lasso by Bayesian prior Laplacian distribution $\mathcal{L}(\mathbf{0}, \lambda_1\mathbf{I})$.
- $TVLR(\gamma)$: VAR by Bayesian prior Gaussian distribution $\mathcal{N}(\mathbf{0}, \gamma^{-1}\mathbf{I}_d)$ and online update with both stationary and drift components of the coefficient [ZWW+16].
- $TVLasso(\lambda_1, \lambda_2)$: VAR-Lasso by Bayesian prior Laplacian distribution $\mathcal{L}(\mathbf{0}, diag(\lambda_1\mathbf{I}, \lambda_2\mathbf{I}))$ and online update with both stationary and drift components of the coefficient [ZWW+16].

Our proposed algorithm, VAR-elastic-net, is denoted as $TVEN(\lambda_{11}, \lambda_{12}, \lambda_{21}, \lambda_{22})$. The penalty parameters λ_{ij} $(i = 1, 2; j = 1, 2)$ are presented in Eq. (10.15), determining the L_1 and L_2 norm of both stationary component and drift component, respectively. During our experiments, we extract small subset of data with early time stamps and employ grid search to find the optimal parameters for all the algorithms. The parameter settings are verified by cross validation in terms of the prediction errors over the extracted data subset.

10.4.2 Evaluation Metrics

- **AUC Score:** At each time t, the AUC score is obtained by comparing its inferred temporal dependency structure with the ground truth. Non-zero value of $W_{l,ji}$ indicates $y_{t-l,i} \rightarrow_g y_{t,j}$ and the higher absolute value of $W_{l,ji}$ implies a larger likelihood of existing a temporal dependency $y_{t-l,i} \rightarrow_g y_{t,j}$.
- **Prediction Error:** At each time t, the true coefficient matrix is W_t and the estimated one is $\hat{W}_t$. Hence, the prediction error ϵ_t defined by the Frobenius norm [CDG00] is $\epsilon_t = \|\hat{W}_t - W_t\|_F$. A smaller prediction error ϵ_t indicates a more accurate inference of the temporal structure.

10.4.3 Synthetic Data and Experiments

In this section, we first present our approach to generate the synthetic data and then illustrate the corresponding experimental results.

Synthetic Data Generation

By generating synthetic MTS with all types of dependency structures, we are able to comprehensively and systematically evaluate the performance of our proposed

Table 10.1 Parameters for synthetic data generation

Name	Description
K	The number of MTS
T	The total length of MTS with time line
L	The maximum time lag for VAR model
n	The number of different value in the case of piecewise constant
S	The sparsity of the spatial–temporal dependency, denoted as the ratio of zero value coefficients in dependency matrix W
μ	The mean of the noise
σ^2	The variance of the noise

method in every scenario. Table 10.1 summarizes the parameters used to generate the synthetic data.

The dependency structure is shown by the coefficient matrix $W_{l,ji}$, which have been constructed by five ways in [ZWW$^+$16], such as **zero value**, **constant value**, **piecewise constant**, **periodic value**, and **random walk**. To show the efficiency of our proposed algorithm, we add a new construct by **grouped value**. The variables are categorized into several groups. Each group first appoints a representative variable whose coefficient is sampled at time t. Meanwhile, the coefficients for other variables at the group are assigned with the same value adding a small Gaussian noise, that is, $\epsilon = 0.1\epsilon^*$, $\epsilon^* \sim \mathcal{N}(0, 1)$.

Overall Evaluation

We first conduct an overall evaluation in terms of AUC and prediction error over synthetic data generated by setting parameter $S = 0.85$, $T = 5000$, $n = 10$, $L = 2$, $\mu = 0$, $\sigma^2 = 1$, and $K = (30, 40, 50)$. From the experimental results shown in Fig. 10.1, we conclude that our proposed method has the best performance in both evaluation metrics, AUC and prediction error, which indicates the superiority of our algorithm in dependency discovery for time-series data.

To better show the capability of our algorithm in capturing the dynamic dependencies, we visualize and compare the ground truth coefficients and the estimated ones by different algorithms over synthetic data with all aforementioned dependency structures. The experiments start with simulations where $S = 0.87$, $T = 3000$, $L = 2$, $n = 10$, $\mu = 0$, and $\sigma^2 = 1$. In order to guarantee consistent comparison with the result in work [ZWW$^+$16], we set parameter $K = 20$. The result shown in Fig. 10.2 indicates that our proposed approach is able to better track the dynamic temporal dependencies in all cases.

Group Effect

To present the ability of our algorithm in better stability and group variable selection, we highlight our experiments on synthetic data with a *group value* dependency structure, where $T = 3000$, $n = 10$, $L = 1$, $\mu = 0$, $\sigma^2 = 1$, and $K = 20$. Among the dependency matrix sampled in this experiment, only 6 coefficients are non-zero and we equally categorize them into two groups. When sampling the coefficient

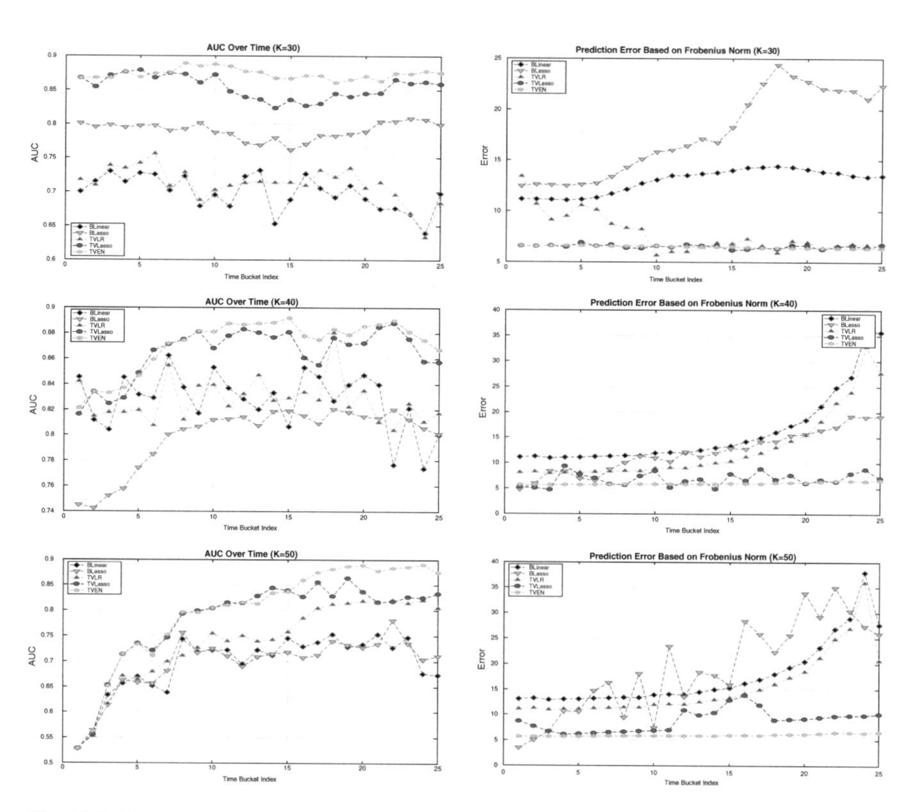

Fig. 10.1 The temporal dependency identification performance is evaluated in terms of AUC and prediction error for algorithms `BLR`(1.0) `BLasso`(1k), `TVLR`(1.0), `TVLasso`(1k, 2k), `TVElastic-Net`(1k, 2k, 1m, 2m). The bucket size is 200

values for each group, we first sample a value x and every member in this group is assigned with x adding a small Gaussian noise, that is, $\epsilon = 0.1\epsilon^*$, $\epsilon^* \sim \mathcal{N}(0, 1)$, such that the synthetic data will have group effect.

We make use of the tuning parameter *shrinkage ratio* s [ZH05] defined as follows:

$$s = \|\beta\|_1 / \max(\|\beta\|_1),$$

where s is a value in $[0, 1]$. A smaller value s indicates a stronger penalty on the L_1 norm of coefficient β, thus a smaller ratio of non-zero coefficient. We also have the following definition:

Definition 10.2 (Zero Point) A zero point for a variable α in our model is equal to the value of *shrinkage ratio* s, which makes the coefficient of the variable α happen to change from zero to non-zero or vice versa.

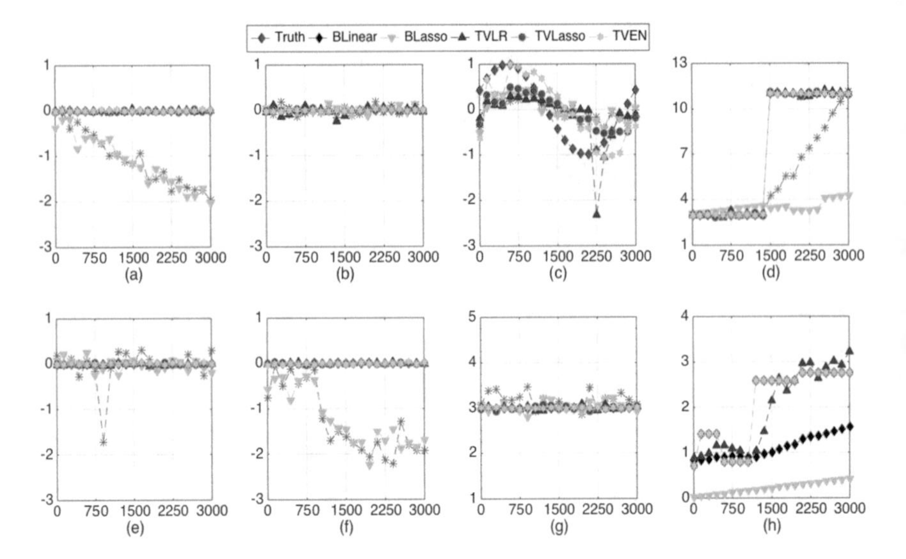

Fig. 10.2 The temporal dependencies from 20 time series are learned and eight coefficients among all are selected for demonstration. Coefficients with zero values are displayed in (**a**), (**b**), (**e**), and (**f**). The coefficients with periodic change, piecewise constant, constant, and random walk are shown in (**c**), (**d**), (**g**), and (**h**), respectively

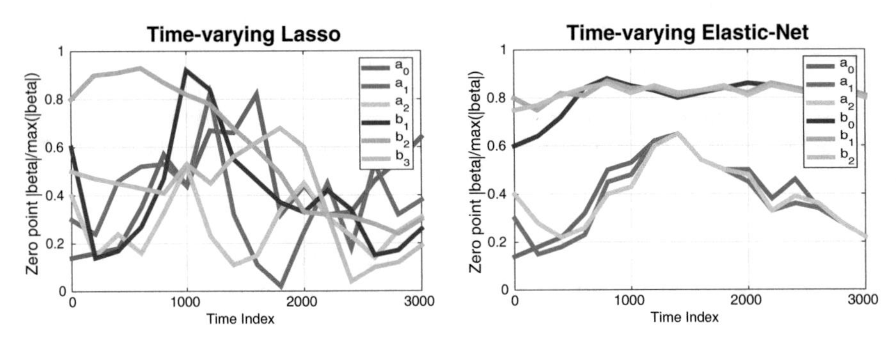

Fig. 10.3 The zero point s changes with time between TVLasso and TVEN. The penalty parameters are $\lambda_1 = \lambda_2 = 1000$ for TVLasso and $\lambda_{11} = \lambda_{12} = \lambda_{21} = \lambda_{22} = 1000$ for TVEN

From the definition of *shrinkage ratio* s, (1) a small zero point for variable α indicates a strong correlation between the variable α and the dependent variable and (2) group variables have closer zero points. However, it is a static result in [ZH05]. Here, we show the dynamic change of zero point with time.

Figure 10.3 records the zero points for all variables with non-zero coefficients calculated by algorithm TVLasso and TVEN. From the result, it is safe to claim that Lasso regularization alone fails to identity group variables; meanwhile, our proposed method with elastic-net regularization succeeds.

10.4.4 Climate Data and Experiments

In this section, we conduct experiments on real-world climate data and display the corresponding experimental results and analysis.

Data Source and Pre-Processing
The MTS data records monthly global surface air temperature from 1948 to 2017 for each grid. The whole global surface is equally segmented into 360×720 grids (0.5 degree latitude $\times$ 0.5 degree longitude global grid for each).

In this paper, we only focus on the East Florida area in the USA and are able to extract totally 46 contiguous land grids with complete monthly air temperature observations from January 1948 to December 2016. Each grid data is considered as one time series $\mathbf{y}$, so the number of multivariate time series K is 46 and the total length of the time series T is 828. Normalization is applied to the data set for all grids.

Spatial–Temporal Overall Evaluation
To illustrate the efficacy of our algorithm on the real-world climate data, we conduct experiments to inspect the prediction performance of our algorithm in the perspective of both space and time.

Figure 10.4 shows the average predicted value of the normalized air temperatures on the total 46 grids of East Florida, where the basic parameters are set to $K = 46$, $T = 828$, $L = 1$, $s = 0.85$, $\mu = 0$, and $\sigma^2 = 1$ for all the algorithms. As illustrated in Fig. 10.4, our algorithm outperforms other baseline algorithms in predicting ability.

Group Effect
In this section, we further conduct experiments on the data of the 11 contiguous grids, a subset of aforementioned 46 points, to illustrate the ability of our algorithm in identifying group locations having similar dependencies towards one another location. Unlike the dependency analysis on the data of the globe or the entire USA [LLNM$^+$09, Cas16], we ignore the influence of the weather in other far regions in our experiments since they are considered insignificant to the 11 grids [KKR$^+$13], a relatively small area. We analyze the dependency matrices of the 11 locations towards two locations $(81.25° W, 27.25° N)$ and $(81.75° W, 30.25° N)$ to show the group effects of the air temperature among those locations.

Figure 10.5 shows the experimental results for the two target locations (black points), respectively. 4 groups are identified among the 11 locations by adjusting the *shrinkage ratio* s and locations in the same group are displayed with same colors. As shown in Fig. 10.5, the black location, i.e., itself, has the most significant correlation for estimating the target air temperatures from time $t-1$ to time t for both the target locations. The relatively close locations in color green and blue have larger power

https://www.esrl.noaa.gov/psd/data/gridded/data.ghcncams.html

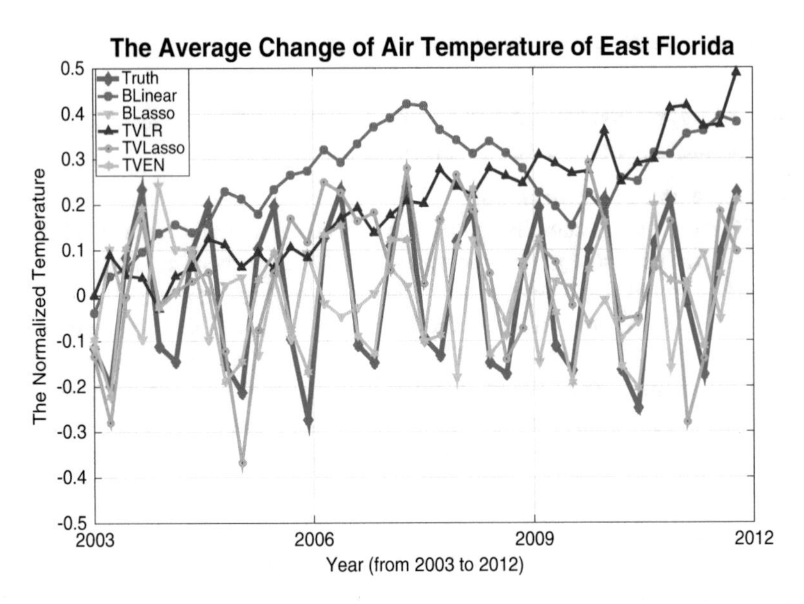

Fig. 10.4 Average predicted value of the normalized air temperature on the total 46 grids of East Florida

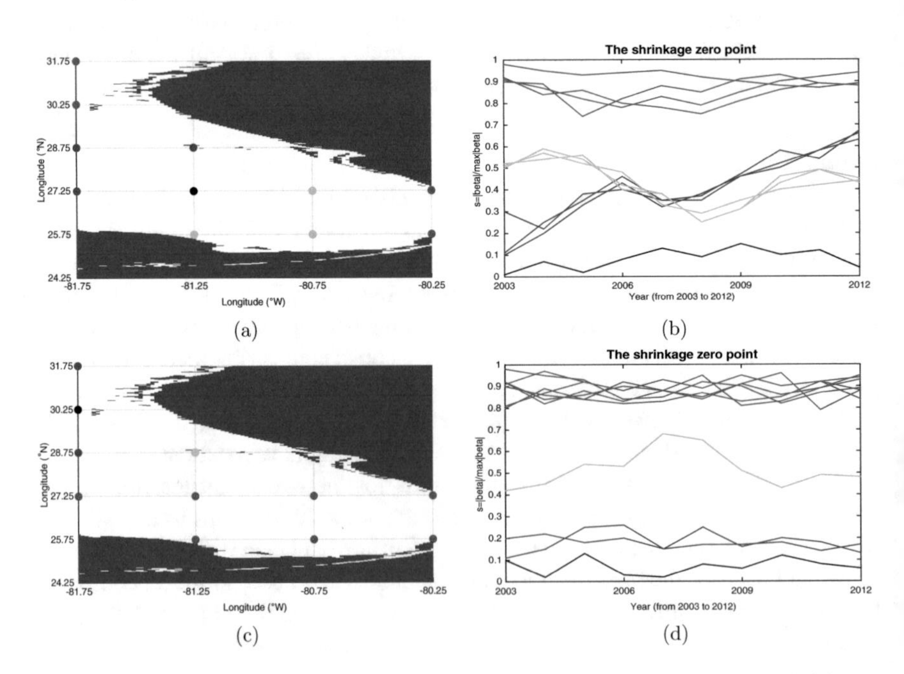

Fig. 10.5 Group dependencies of air temperatures over time for two target locations. Subfigures (**a**) and (**c**) show the geographical locations and target locations are in black. Subfigures (**b**) and (**d**) show the zero points graph for the two target locations, respectively

for predicting the target air temperatures from time $t - 1$ to time t for the two locations than the red location.

The spatial–temporal dependency structures learned in our experiments are quite consistent with domain expertise which indicates our model is able to provide significant insights in MTS data.

10.5 Conclusion and Future Work

In this chapter, we proposed a novel VAR-elastic-net model with online Bayesian update allowing for both stable-sparsity and group selection among MTS, which implements adaptive inference strategy of particle learning. Extensive empirical studies on both the synthetic and real MTS data demonstrate the effectiveness and the efficiency of the proposed method.

In the process of time-varying temporal dependency discovery from MTS, the choice of regularizer is essential. One possible future work is to automate the identification of the proper regularizer for different MTS in an online setting. Another possible direction is to apply other dimension deduction tool, such as principal component analysis, that extracts the characters on the dynamical process. Finally, the dynamical structure can be also improved, such as particle learning or Gauss random walk, to propose the dynamical model to simulate the real phenomenon.

Chapter 11
Conclusion

Now more than ever machine learning and embedded AI will be essential in maintaining information assurance for all aspects of our nation's security and defense, as well as every transaction we make in government and commercial or private business operations. Information is the lifeblood of every business transaction and managing risks related to the use, processing, storage, analysis, and transmission of this information, as well as the enormous "big data" analytics systems upon which we now rely are the vital parts allowing us to process and sustain the flow of information for our modern world. As we increase the numbers of devices and interconnected networks, especially as the Internet of things begins to blossom, enormous risks will continue to emerge. Any disruption to our systems and processes could cause economic collapse for a business, as well as our nation.

This book represents a major contribution in terms of mathematical aspects of machine learning by the authors and collaborators. What we have tried to portray in this book is the current state of the art for machine learning and associated artificial intelligence techniques. The algorithms presented here have been designed to find the local minima in convex optimization schemes and to obtain frictionless global minima from Newton's second law. We believe we have provided a solid theoretical framework upon which further analysis and research can be conducted. We hope this book has been beneficial to you in helping to identify and address existing issues in the fields of machine learning, artificial intelligence, deep neural networks, as well as a plethora of emerging fields. By highlighting a few popular techniques, and demonstrating our new CoCoSSC methodology, to resolve the noisy subspace clustering challenge, we have provided what we consider to be a significant improvement and more robust solution than current methodologies provide. Our numerical results confirm the effectiveness and efficiency of this new method, which we hope will provide a springboard to enhanced operations in the many fields in which it expected to be used. More importantly we hope this new methodology provides a deeper understanding for researchers, as they take this work to the next level.

B. Shi, S. S. Iyengar, *Mathematical Theories of Machine Learning - Theory and Applications*, https://doi.org/10.1007/978-3-030-17076-9_11

References

[AAB+17] N. Agarwal, Z. Allen-Zhu, B. Bullins, E. Hazan, T. Ma, Finding approximate local minima faster than gradient descent, in *STOC* (2017), pp. 1195–1199. http://arxiv.org/abs/1611.01146

[AAZB+17] N. Agarwal, Z. Allen-Zhu, B. Bullins, E. Hazan, T. Ma, Finding approximate local minima faster than gradient descent, in *Proceedings of the 49th Annual ACM SIGACT Symposium on Theory of Computing* (ACM, New York, 2017), pp. 1195–1199

[AG16] A. Anandkumar, R. Ge, Efficient approaches for escaping higher order saddle points in non-convex optimization, in *Conference on Learning Theory* (2016), pp. 81–102. arXiv preprint arXiv:1602.05908

[ALA07] A. Arnold, Y. Liu, N. Abe, Temporal causal modeling with graphical granger methods, in *Proceedings of the 13th ACM SIGKDD International Conference on Knowledge Discovery and Data Mining* (ACM, New York, 2007), pp. 66–75

[B+15] S. Bubeck, Convex optimization: algorithms and complexity. Found. Trends in Mach. Learn. **8**(3–4), 231–357 (2015)

[BBW+90] F.P. Bretherton, K. Bryan, J.D. Woods et al., Time-dependent greenhouse-gas-induced climate change. Clim. Change IPCC Sci. Assess. **1990**, 173–194 (1990)

[BJ03] R. Basri, D. Jacobs, Lambertian reflectance and linear subspaces. IEEE Trans. Pattern Anal. Mach. Intell. **25**(2), 218–233 (2003)

[BJRL15] G.E.P. Box, G.M. Jenkins, G.C. Reinsel, G.M. Ljung, *Time Series Analysis: Forecasting and Control* (Wiley, London, 2015)

[BL12] M.T. Bahadori, Y. Liu, On causality inference in time series, in *AAAI Fall Symposium: Discovery Informatics* (2012)

[BL13] M.T. Bahadori, Y. Liu, An examination of practical granger causality inference, in *Proceedings of the 2013 SIAM International Conference on data Mining* (SIAM, 2013), pp. 467–475

[BLE17] S. Bubeck, Y.T. Lee, R. Eldan, Kernel-based methods for bandit convex optimization, in *Proceedings of the 49th Annual ACM SIGACT Symposium on Theory of Computing* (ACM, New York, 2017), pp. 72–85

[BM00] P.S. Bradley, O.L. Mangasarian, *K*-plane clustering. J. Global Optim. **16**(1), 23–32 (2000)

[BNS16] S. Bhojanapalli, B. Neyshabur, N. Srebro, Global optimality of local search for low rank matrix recovery, in *Advances in Neural Information Processing Systems* (2016), pp. 3873–3881

B. Shi, S. S. Iyengar, *Mathematical Theories of Machine Learning - Theory and Applications*, https://doi.org/10.1007/978-3-030-17076-9

[BPC+11] S. Boyd, N. Parikh, E. Chu, B. Peleato, J. Eckstein, Distributed optimization and statistical learning via the alternating direction method of multipliers. Found. Trends Mach. Learn. **3**(1), 1–122 (2011)

[BT09] A. Beck, M. Teboulle, A fast iterative shrinkage-thresholding algorithm for linear inverse problems. SIAM J. Imag. Sci. **2**(1), 183–202 (2009)

[BV04] S. Boyd, L. Vandenberghe, *Convex Optimization* (Cambridge University Press, Cambridge, 2004)

[Cas16] S. Castruccio, Assessing the spatio-temporal structure of annual and seasonal surface temperature for CMIP5 and reanalysis. Spatial Stat. **18**, 179–193 (2016)

[CD16] Y. Carmon, J.C. Duchi, Gradient descent efficiently finds the cubic-regularized non-convex Newton step. arXiv preprint arXiv:1612.00547 (2016)

[CDG00] B. Carpentieri, I.S. Duff, L. Giraud, Sparse pattern selection strategies for robust frobenius-norm minimization preconditioners in electromagnetism. Numer. Linear Algebr. Appl. **7**(7–8), 667–685 (2000)

[CDHS16] Y. Carmon, J.C. Duchi, O. Hinder, A. Sidford, Accelerated methods for non-convex optimization. arXiv preprint arXiv:1611.00756 (2016)

[CJLP10] C.M. Carvalho, M.S. Johannes, H.F. Lopes, N.G. Polson, Particle learning and smoothing. Stat. Sci. **25**, 88–106 (2010)

[CJW17] Z. Charles, A. Jalali, R. Willett, Sparse subspace clustering with missing and corrupted data. arXiv preprint: arXiv:1707.02461 (2017)

[CK98] J. Costeira, T. Kanade, A multibody factorization method for independently moving objects. Int. J. Comput. Vis. **29**(3), 159–179 (1998)

[CLLC10] X. Chen, Y. Liu, H. Liu, J.G. Carbonell, Learning spatial-temporal varying graphs with applications to climate data analysis, in *AAAI* (2010)

[CR09] E.J. Candès, B. Recht, Exact matrix completion via convex optimization. Found. Comput. Math. **9**(6), 717–772 (2009)

[CRS14] F.E. Curtis, D.P. Robinson, M. Samadi, A trust region algorithm with a worst-case iteration complexity of $O(\epsilon^{-3/2})$ for nonconvex optimization. Math. Program. **162**(1–2), 1–32 (2014)

[CT05] E.J. Candes, T. Tao, Decoding by linear programming. IEEE Trans. Inf. Theory **51**(12), 4203–4215 (2005)

[CT07] E. Candes, T. Tao, The Dantzig selector: statistical estimation when p is much larger than n. Ann. Stat. **35**(6), 2313–2351 (2007)

[DGA00] A. Doucet, S. Godsill, C. Andrieu, On sequential Monte Carlo sampling methods for bayesian filtering. Stat. Comput. **10**(3), 197–208 (2000)

[DJL+17] S.S. Du, C. Jin, J.D. Lee, M.I. Jordan, B. Poczos, A. Singh, Gradient descent can take exponential time to escape saddle points, in *Proceedings of Advances in Neural Information Processing Systems (NIPS)* (2017), pp. 1067–1077

[DKZ+03] P.M. Djuric, J.H. Kotecha, J. Zhang, Y. Huang, T. Ghirmai, M.F. Bugallo, J. Miguez, Particle filtering. IEEE Signal Process. Mag. **20**(5), 19–38 (2003)

[DZ17] A. Datta, H. Zou, Cocolasso for high-dimensional error-in-variables regression. Ann. Stat. **45**(6), 2400–2426 (2017)

[EBN12] B. Eriksson, L. Balzano, R. Nowak, High rank matrix completion, in *Artificial Intelligence and Statistics* (2012), pp. 373–381

[Eic06] M. Eichler, Graphical modelling of multivariate time series with latent variables. Preprint, Universiteit Maastricht (2006)

[EV13] E. Elhamifar, R. Vidal, Sparse subspace clustering: algorithm, theory, and applications. IEEE Trans. Pattern Anal. Mach. Intell. **35**(11), 2765–2781 (2013)

[FKM05] A.D. Flaxman, A.T. Kalai, H.B. McMahan, Online convex optimization in the bandit setting: gradient descent without a gradient, in *Proceedings of the Sixteenth Annual ACM-SIAM Symposium on Discrete Algorithms* (Society for Industrial and Applied Mathematics, Philadelphia, 2005), pp. 385–394

[FM83] E.B. Fowlkes, C.L. Mallows, A method for comparing two hierarchical clusterings. J. Am. Stat. Assoc. **78**(383), 553–569 (1983)

[GHJY15] R. Ge, F. Huang, C. Jin, Y. Yuan, Escaping from saddle points—online stochastic gradient for tensor decomposition, in *Proceedings of the 28th Conference on Learning Theory* (2015), pp. 797–842

[GJZ17] R. Ge, C. Jin, Y. Zheng, No spurious local minima in nonconvex low rank problems: a unified geometric analysis, in *Proceedings of the 34th International Conference on Machine Learning* (2017), pp. 1233–1242

[GLM16] R. Ge, J.D. Lee, T. Ma, Matrix completion has no spurious local minimum, in *Advances in Neural Information Processing Systems* (2016), pp. 2973–2981

[GM74] P.E. Gill, W. Murray, Newton-type methods for unconstrained and linearly constrained optimization. Math. Program. **7**(1), 311–350 (1974)

[Gra69] C.W.J. Granger, Investigating causal relations by econometric models and cross-spectral methods. Econometrica **37**(3), 424–438 (1969)

[Gra80] C.W.J. Granger, Testing for causality: a personal viewpoint. J. Econ. Dyn. Control. **2**, 329–352 (1980)

[Ham94] J.D. Hamilton, *Time Series Analysis*, vol. 2 (Princeton University Press, Princeton, 1994)

[Har71] P. Hartman, The stable manifold of a point of a hyperbolic map of a banach space. J. Differ. Equ. **9**(2), 360–379 (1971)

[Har82] P. Hartman, *Ordinary Differential Equations, Classics in Applied Mathematics*, vol. 38 (Society for Industrial and Applied Mathematics (SIAM), Philadelphia, 2002). Corrected reprint of the second (1982) edition 1982

[Har90] A.C. Harvey, *Forecasting, Structural Time Series Models and the Kalman Filter* (Cambridge University Press, Cambridge, 1990)

[HB15] R. Heckel, H. Bölcskei, Robust subspace clustering via thresholding. IEEE Trans. Inf. Theory **61**(11), 6320–6342 (2015)

[Hec98] D. Heckerman, A tutorial on learning with bayesian networks. Learning in Graphical Models (Springer, Berlin, 1998), pp. 301–354

[HL14] E. Hazan, K. Levy, Bandit convex optimization: towards tight bounds, in *Advances in Neural Information Processing Systems* (2014), pp. 784–792

[HMR16] M. Hardt, T. Ma, B. Recht, Gradient descent learns linear dynamical systems. arXiv preprint arXiv:1609.05191 (2016)

[HTB17] R. Heckel, M. Tschannen, H. Bölcskei, Dimensionality-reduced subspace clustering. Inf. Inference: A J. IMA **6**(3), 246–283 (2017)

[JCSX11] A. Jalali, Y. Chen, S. Sanghavi, H. Xu, *Clustering Partially Observed Graphs Via Convex Optimization* (ICML, 2011)

[JGN+17] C. Jin, R. Ge, P. Netrapalli, S.M. Kakade, M.I. Jordan, How to escape saddle points efficiently, in *Proceedings of the 34th International Conference on Machine Learning* (2017), pp. 1724–1732

[JHS+11] M. Joshi, E. Hawkins, R. Sutton, J. Lowe, D. Frame, Projections of when temperature change will exceed 2 [deg] c above pre-industrial levels. Nat. Clim. Change **1**(8), 407–412 (2011)

[JNJ17] C. Jin, P. Netrapalli, M.I. Jordan, Accelerated gradient descent escapes saddle points faster than gradient descent. arXiv preprint arXiv:1711.10456 (2017)

[JYG+03] R. Jansen, H. Yu, D. Greenbaum, Y. Kluger, N.J. Krogan, S. Chung, A. Emili, M. Snyder, J.F. Greenblatt, M. Gerstein, A bayesian networks approach for predicting protein–protein interactions from genomic data. Science **302**(5644), 449–453 (2003)

[KKR+13] W. Kleiber, R.W. Katz, B. Rajagopalan et al., Daily minimum and maximum temperature simulation over complex terrain. Ann. Appl. Stat. **7**(1), 588–612 (2013)

[KMO10] R.H. Keshavan, A. Montanari, S. Oh, Matrix completion from a few entries. IEEE Trans. Inf. Theory **56**(6), 2980–2998 (2010)

[LBL12] Y. Liu, T. Bahadori, H. Li, Sparse-GEV: sparse latent space model for multivariate extreme value time serie modeling. arXiv preprint arXiv:1206.4685 (2012)

[LKJ09] Y. Liu, J.R. Kalagnanam, O. Johnsen, Learning dynamic temporal graphs for oil-production equipment monitoring system, in *Proceedings of the 15th ACM SIGKDD International Conference on Knowledge Discovery and Data Mining* (ACM, New York, 2009), pp. 1225–1234

[LL+10] Q. Li, N. Lin, The Bayesian elastic net. Bayesian Anal. **5**(1), 151–170 (2010)

[LLNM+09] A.C. Lozano, H. Li, A. Niculescu-Mizil, Y. Liu, C. Perlich, J. Hosking, N. Abe, Spatial-temporal causal modeling for climate change attribution, in *Proceedings of the 15th ACM SIGKDD international conference on Knowledge discovery and data mining* (ACM, New York, 2009), pp. 587–596

[LLY+13] G. Liu, Z. Lin, S. Yan, J. Sun, Y. Yu, Y. Ma, Robust recovery of subspace structures by low-rank representation. IEEE Trans. Pattern Anal. Mach. Intell. **35**(1), 171–184 (2013)

[LNMLL10] Y. Liu, A. Niculescu-Mizil, A.C. Lozano, Y. Lu, Learning temporal causal graphs for relational time-series analysis, in *Proceedings of the 27th International Conference on Machine Learning (ICML-10)* (2010), pp. 687–694

[LPP+17] J.D. Lee, I. Panageas, G. Piliouras, M. Simchowitz, M.I. Jordan, B. Recht, First-order methods almost always avoid saddle points. arXiv preprint arXiv:1710.07406 (2017)

[LRP16] L. Lessard, B. Recht, A. Packard, Analysis and design of optimization algorithms via integral quadratic constraints. SIAM J. Optim. **26**(1), 57–95 (2016)

[LSJR16] J.D. Lee, M. Simchowitz, M.I. Jordan, B. Recht, Gradient descent only converges to minimizers, in *Conference on Learning Theory* (2016), pp. 1246–1257

[LWL+16] X. Li, Z. Wang, J. Lu, R. Arora, J. Haupt, H. Liu, T. Zhao, Symmetry, saddle points, and global geometry of nonconvex matrix factorization. arXiv preprint arXiv:1612.09296 (2016)

[LY+84] D.G. Luenberger, Y. Ye et al., *Linear and Nonlinear Programming*, vol. 2 (Springer, Berlin, 1984)

[LY17] M. Liu, T. Yang, On noisy negative curvature descent: competing with gradient descent for faster non-convex optimization. arXiv preprint arXiv:1709.08571 (2017)

[LZZ+16] T. Li, W. Zhou, C. Zeng, Q. Wang, Q. Zhou, D. Wang, J. Xu, Y. Huang, W. Wang, M. Zhang et al., DI-DAP: an efficient disaster information delivery and analysis platform in disaster management, in *Proceedings of the 25th ACM International on Conference on Information and Knowledge Management* (ACM, New York, 2016), pp. 1593–1602

[MDHW07] Y. Ma, H. Derksen, W. Hong, J. Wright, Segmentation of multivariate mixed data via lossy data coding and compression. IEEE Trans. Pattern Anal. Mach. Intell. **29**(9), 1546–1562 (2007)

[MS79] J.J. Moré, D.C. Sorensen, On the use of directions of negative curvature in a modified newton method. Math. Program. **16**(1), 1–20 (1979)

[Mur02] K.P. Murphy, Dynamic bayesian networks: representation, inference and learning, Ph.D. thesis, University of California, Berkeley, 2002

[Mur12] K.P. Murphy, *Machine Learning: A Probabilistic Perspective* (MIT Press, Cambridge, MA, 2012)

[Nes83] Y. Nesterov, *A Method of Solving a Convex Programming Problem with Convergence Rate o (1/k2)* Soviet Mathematics Doklady, vol. 27 (1983), pp. 372–376

[Nes13] Y. Nesterov, *Introductory Lectures on Convex Optimization: A Basic Course*, vol. 87 (Springer, Berlin, 2013)

[NH11] B. Nasihatkon, R. Hartley, Graph connectivity in sparse subspace clustering, in *CVPR* (IEEE, Piscataway, 2011)

[NN88] Y. Nesterov, A. Nemirovsky, A general approach to polynomial-time algorithms design for convex programming, Tech. report, Technical report, Centr. Econ. & Math. Inst., USSR Acad. Sci., Moscow, USSR, 1988

[NP06] Y. Nesterov, B.T. Polyak, Cubic regularization of newton method and its global performance. Math. Program. **108**(1), 177–205 (2006)

[OC15] B. O'Donoghue, E. Candès, Adaptive restart for accelerated gradient schemes. Found. Comput. Math. **15**(3), 715–732 (2015)

[OW17] M. O'Neill, S.J. Wright, Behavior of accelerated gradient methods near critical points of nonconvex problems. arXiv preprint arXiv:1706.07993 (2017)

[PCS14] D. Park, C. Caramanis, S. Sanghavi, Greedy subspace clustering, in *Advances in Neural Information Processing Systems* (2014), pp. 2753–2761

[Pem90] R. Pemantle, Nonconvergence to unstable points in urn models and stochastic approximations. Ann. Probab. **18**(2), 698–712 (1990)

[Per13] L. Perko, *Differential Equations and Dynamical Systems*, vol. 7 (Springer, Berlin, 2013)

[PKCS17] D. Park, A. Kyrillidis, C. Carmanis, S. Sanghavi, Non-square matrix sensing without spurious local minima via the Burer-Monteiro approach, in *Proceedings of the 20th International Conference on Artificial Intelligence and Statistics* (2017), pp. 65–74

[Pol64] B.T. Polyak, Some methods of speeding up the convergence of iteration methods. USSR Comput. Math. Math. Phys. **4**(5), 1–17 (1964)

[Pol87] B.T. Polyak, *Introduction to Optimization* (Translations series in mathematics and engineering) (Optimization Software, 1987)

[PP16] I. Panageas, G. Piliouras, Gradient descent only converges to minimizers: non-isolated critical points and invariant regions. arXiv preprint arXiv:1605.00405 (2016)

[QX15] C. Qu, H. Xu, Subspace clustering with irrelevant features via robust dantzig selector, in *Advances in Neural Information Processing Systems* (2015), pp. 757–765

[Rec11] B. Recht, A simpler approach to matrix completion. J. Mach. Learn. Res. **12**, 3413–3430 (2011)

[RHW$^+$88] D.E. Rumelhart, G.E. Hinton, R.J. Williams et al., Learning representations by back-propagating errors. Cogn. Model. **5**(3), 1 (1988)

[RS15] V.K. Rohatgi, A.K.M.E. Saleh, *An Introduction to Probability and Statistics* (Wiley, London, 2015)

[RW17] C.W. Royer, S.J. Wright, Complexity analysis of second-order line-search algorithms for smooth nonconvex optimization. arXiv preprint arXiv:1706.03131 (2017)

[RZS$^+$17] S.J. Reddi, M. Zaheer, S. Sra, B. Poczos, F. Bach, R. Salakhutdinov, A.J. Smola, A generic approach for escaping saddle points. arXiv preprint arXiv:1709.01434 (2017)

[SBC14] W. Su, S. Boyd, E. Candes, A differential equation for modeling Nesterov's accelerated gradient method: theory and insights, in *Advances in Neural Information Processing Systems* (2014), pp. 2510–2518

[SC12] M. Soltanolkotabi, E.J. Candes, A geometric analysis of subspace clustering with outliers. Ann. Stat. **40**(4), 2195–2238 (2012)

[SEC14] M. Soltanolkotabi, E. Elhamifar, E.J. Candes, Robust subspace clustering. Ann. Stat. **42**(2), 669–699 (2014)

[SHB16] Y. Shen, B. Han, E. Braverman, Stability of the elastic net estimator. J. Complexity **32**(1), 20–39 (2016)

[Shu13] M. Shub, *Global Stability of Dynamical Systems* (Springer, Berlin, 2013)

[SMDH13] I. Sutskever, J. Martens, G. Dahl, G. Hinton, On the importance of initialization and momentum in deep learning, in *International Conference on Machine Learning* (2013), pp. 1139–1147

[Sol14] M. Soltanolkotabi, Algorithms and theory for clustering and nonconvex quadratic programming. Ph.D. thesis, Stanford University, 2014

[SQW16] J. Sun, Q. Qu, J. Wright, A geometric analysis of phase retrieval, in *2016 IEEE International Symposium on Information Theory (ISIT)* (IEEE, Piscataway, 2016), pp. 2379–2383

[SQW17] J. Sun, Q. Qu, J. Wright, Complete dictionary recovery over the sphere I: overview and the geometric picture. IEEE Trans. Inf. Theory **63**(2), 853–884 (2017)

[TG17] P.A. Traganitis, G.B. Giannakis, Sketched subspace clustering. IEEE Trans. Signal Process. **66**(7), 1663–1675 (2017)

[Tib96] R. Tibshirani, Regression shrinkage and selection via the lasso. J. R. Stat. Soc. Ser. B Methodol. **58**(1), 267–288 (1996)

[TPGC] T. Park, G. Casella, The Bayesian Lasso. J. Am. Stat. Assoc. **103**(482), 681–686 (2008)

[Tse00] P. Tseng, Nearest q-flat to m points. J. Optim. Theory Appl. **105**(1), 249–252 (2000)

[TV17] M. Tsakiris, R. Vidal, Algebraic clustering of affine subspaces. IEEE Trans. Pattern Anal. Mach. Intell. **40**(2), 482–489 (2017)

[TV18] M.C. Tsakiris, R. Vidal, Theoretical analysis of sparse subspace clustering with missing entries. arXiv preprint arXiv:1801.00393 (2018)

[Vid11] R. Vidal, Subspace clustering. IEEE Signal Process. Mag. **28**(2), 52–68 (2011)

[VMS05] R. Vidal, Y. Ma, S. Sastry, Generalized principal component analysis (GPCA). IEEE Trans. Pattern Anal. Mach. Intell. **27**(12), 1945–1959 (2005)

[WN99] S. Wright, J. Nocedal, *Numerical Optimization*, vol. 35, 7th edn. (Springer, Berlin, 1999), pp. 67–68.

[WRJ16] A.C. Wilson, B. Recht, M.I. Jordan, A lyapunov analysis of momentum methods in optimization. arXiv preprint arXiv:1611.02635 (2016)

[WWBS17] Y. Wang, J. Wang, S. Balakrishnan, A. Singh, Rate optimal estimation and confidence intervals for high-dimensional regression with missing covariates. arXiv preprint arXiv:1702.02686 (2017)

[WWJ16] A. Wibisono, A.C. Wilson, M.I. Jordan, A variational perspective on accelerated methods in optimization. Proc. Nat. Acad. Sci. **113**(47), E7351–E7358 (2016)

[WWS15a] Y. Wang, Y.-X. Wang, A. Singh, A deterministic analysis of noisy sparse subspace clustering for dimensionality-reduced data, in *International Conference on Machine Learning* (2015), pp. 1422–1431

[WWS15b] Y. Wang, Y.-X. Wang, A. Singh, Differentially private subspace clustering, in *Advances in Neural Information Processing Systems* (2015), pp. 1000–1008

[WWS16] Y. Wang, Y.-X. Wang, A. Singh, Graph connectivity in noisy sparse subspace clustering, in *Artificial Intelligence and Statistics* (2016), pp. 538–546

[WX16] Y.-X. Wang, H. Xu, Noisy sparse subspace clustering. J. Mach. Learn. Res. **17**(12), 1–41 (2016)

[YP06] J. Yan, M. Pollefeys, A general framework for motion segmentation: independent, articulated, rigid, non-rigid, degenerate and non-degenerate, in *European Conference on Computer Vision* (Springer, Berlin, 2006), pp. 94–106

[YRV15] C. Yang, D. Robinson, R. Vidal, Sparse subspace clustering with missing entries, in *International Conference on Machine Learning* (2015), pp. 2463–2472

[ZF09] C. Zou, J. Feng, Granger causality vs. dynamic bayesian network inference: a comparative study. BMC Bioinf. **10**(1), 122 (2009)

[ZFIM12] A. Zhang, N. Fawaz, S. Ioannidis, A. Montanari, Guess who rated this movie: identifying users through subspace clustering. arXiv preprint arXiv:1208.1544 (2012)

[ZH05] H. Zou, T. Hastie, Regularization and variable selection via the elastic net. J. R. Stat. Soc. Ser. B Stat Methodol. **67**(2), 301–320 (2005)

[ZWML16] C. Zeng, Q. Wang, S. Mokhtari, T. Li, Online context-aware recommendation with time varying multi-armed bandit, in *Proceedings of the 22nd ACM SIGKDD International Conference on Knowledge Discovery and Data Mining* (ACM, New York, 2016), pp. 2025–2034

[ZWW$^+$16] C. Zeng, Q. Wang, W. Wang, T. Li, L. Shwartz, Online inference for time-varying temporal dependency discovery from time series, in *2016 IEEE International Conference on Big Data (Big Data)* (IEEE, Piscataway, 2016), pp. 1281–1290

Publication List

[1] B. Shi, A mathematical framework on machine learning: theory and application, Ph.D dissertation, Florida International University, Miami, Florida, Dec 2018
[2] B. Shi, T. Li, S.S. Iyengar, A conservation law method in optimization, in *10th NIPS Workshop on Optimization for Machine Learning* (2017)
[3] B. Shi, S.S. Du, Y. Wang, J. Lee, Gradient descent converges to minimizers: optimal and adaptive step size rules, in INFORMS J. Optim. (2018). (accepted with minor revision)
[4] W. Wang, B. Shi, T. Li, S.S. Iyengar, Online discovery for stable and grouping causalities in multivariate time series (2018). (submit to TKDD)
[5] Y. Wang, B. Shi, Y. Wang, Y. Tao, S.S. Iyengar, Improved sample complexity in sparse subspace clustering with noisy and missing observations (2018). (submit to AISTATS 2018, joint work with CMU)

B. Shi, S. S. Iyengar, *Mathematical Theories of Machine Learning - Theory and Applications*, https://doi.org/10.1007/978-3-030-17076-9

Index

B. Shi, S. S. Iyengar, *Mathematical Theories of Machine Learning - Theory and Applications*, https://doi.org/10.1007/978-3-030-17076-9

Printed in the United States
By Bookmasters